Stories From The Jeweled Crossing

VOL: 1

The Beginning

By:

Brian James Hess

Copyright © 2015 Brian J Hess

Editorial Provisions made by Mrs. Tammy S Hess

All rights reserved.

ISBN-13:978-1517508081

To all of you who take the time

to read this book;

What you should know is that the stories about me are true. The names I have changed on some of the people I have known because some of the things we did where illegal.

The facts of some have been softened but I think that you will get the idea of what was going on in my mind and in the stories.

The stories of the people I have known I believe where told to me with all the truth they could speak of.

They had no reason to lie to me, they wanted me to know what happened to them, they trusted me and most of them still do. They just wanted someone to know.

I wrote it in a simple way, there were very few words I had to look up in the dictionary.

The people in this book seem to be what some people call simple people but they are so far from the truth of these people it's funny.

Some don't read well and I want them to read what I have wrote; I also know that there are others out there who have stories of their own and I want them to be able to relate to ours. The K.I.S.S. method seems to work best for us writing this book.

Writing this book brought back a lot of memories that were bad and some tore the soul right out of me.

I went out of my mind for a while as I searched for the pieces scattered all around, vultures eating pieces of it. Some I know I will never get back.

There was also a lot of exciting times. I have seen things and been in places where a lot of people only dream of.

Some places money won't buy you in. It was about who you know, other times it was just about how much you were willing to risk.

In these places you see life raw. It taught me to appreciate the simple things in life and how much it means to a human being to have someone to trust without a doubt.

When you are on the path you believe you're supposed to be on it takes that trust and their in-put. They ARE a major key in the pursuit of your goal that you are after.

It's the kind money can't buy or the kind the mass media preaches on T.V. You will not find it on your home computer either. It's the kind of truth drugs and alcohol won't hide or erase.

You have to believe, have to know is a better way of explaining it

My father would say don't listen to the Ney Sayers, that there is no time for them, you can't help them, they don't understand,

YOU CAN'T GIVE EM' WHAT IT IS AND THEIR NOT LOOKING FOR IT SO MOVE ON AND LOOK FOR THOSE WHO WANT TO KNOW.

MY FATHER IS A BIG PART OF WHO I AM, SO IS MY GRANDFATHER. THEY KNEW AND TAUGHT IT TO ME. SO I PRESSED FORWARD ALWAYS LOOKING.

SOME DAYS ARE GREAT AND SOME ARE OKAY. I LIKE THEM. THE OTHER DAYS I JUST DEAL WITH ONE MOMENT AT A TIME.

MAYBE THERE IS SOMETHING THAT YOU CAN TAKE AWAY FROM THIS BOOK. ONE THING RIGHT UP FRONT IN MY MIND IS THEY ALL HAVE LOST THEIR MINDS.

* MAYBE SO, I DO HOWEVER BELIEVE THAT I HAVE BEEN TRULY BLESSED*

THERE IS NO OTHER EXPLANATION OF HOW MY FATHER AND THE OTHER PEOPLE I KNOW MADE IT THROUGH SOME OF THE STUFF LIFE THREW AT THEM. SOME THEY COULDN'T CONTROL, OTHERS, JUST BAD TIMING.

"THANK GOD FOR EVERY GOOD DAY"

"THANK GOD FOR HELPING YOU THROUGH EVERY BAD DAY"

IT MAY BE THAT THERE IS A LESSON TO BE LEARNED.

❖

❖

IF YOU ARE NOT A BELIEVER IN GOD, ASK WHO YOU BELIEVE IN TO BLESS YOU WITH A LIFE THAT IS FULL. THERE ARE THOSE WHO NEVER GET THE CHANCE TO TRULY LIVE FOR ONE REASON OR ANOTHER.

❖

❖

PLEASE:

READ ON

ABOUT THE AUTHOR

There's nothing for me to tell you that you won't read about in my books. I'll let it all hang-out, Good or Bad!

Yes I said Book's. There are a lot of stories and not just the ones I have told sitting around a burning Alabama Fire-Log. Trust me: I've burned a lot of them.

I really didn't write them for any one person. I wrote them because I believe it is something that I was supposed to do. Reading my books may make that last sentence make sense.

I have had a lot of adventures in my life or maybe it just seemed that way to me. Some people just think I am crazy; So be it.

One thing's for sure, I was given the time and all the material I needed to write this book There's really nothing more to say. I'm sure you can figure it out.

My Thank You List:

My Thank You list is not that long. I thank the Lord for every day I have been given. There are days I still do not understand; And in that, My Thank You list for days given is never ending. I don't know where to start but I will say that that's between who and what I believe in.

There is someone else I would like to Thank. She is the only one who could have help me put this together. She has a special Gift. I know that for a fact!

When I was talking to her about this adventure she never hesitated. All the hard work, trying to make sense of what I had put on paper, then listening to me tell her how I wanted it to look on print.

She just smiled and said "Okay". It should be this way. I wanted it my way and she just said "Okay".

She is my current wife, Tammy.

She told me not to say "Third Wife" because it sounded like there would be more. She laughed then.

I don't know how she puts up with me on some days. She's just another blessing to me. There will be a big story behind Tammy and myself but that's for later.

Bottom-Line; We put what you're going to read together and we hope you like it. There's more to come. Maybe I will listen to her a little more:

I LOVE HER

SHE'S MY FRIEND AS WELL AS MY WIFE

ROLL THE DICE TAMMY

"LET'S DO IT"

DRUG FROM THE FIRE BY YOU;

> YOUR' CURRENT HUSBAND;
> BRIAN HESS

TABLE OF CONTENTS:

- * Title Page — i
- * Introduction — iv
- * About the Author — ix
- * My Thank You List — x

1. As It Was — Pg 1
2. Charlie — Pg 11
3. The Farm House — Pg 25
4. Back On The Road — Pg 107
5. It Begins — Pg 181

- * Dictionary — Pg 285

Part One:

As It Was

Standing beside my car that I had just spent the night in, I began to stretch my body. I was all stoved up from sleeping leaned over the steering wheel. Stretching my arms and tilting my head back I noticed the sky was a beautiful shade of blue and the sun was warm on my face. There were a few clouds. They looked like cotton balls just floating there.

At that moment I must have moved my feet because I just about fell to the ground. Looking down I realized I was standing in this red, wet and slimy clay like dirt. Just off the side of the road the ground had a slight slope toward a run off ditch.

It must have rained a little while I slept.

The ground around me was slightly wet. It was summer time and it rains around here at night to cool things off I was later informed.

Putting my hands on the fender of my car to get my balance back the next rude awakening hit me right in the nose. It was an awful smell.

It took me a minute to realize the smell was an after effect of a serpentine belt on the motor which caused my car to overheat, leaving me to spend the night, sweating, leaned over the steering wheel.

Chuckling to myself at the predicament I was in, I walked a few steps up to the black top.

Looking first to my left, that was the way I had traveled to get to where I am at; there were trees on both sides of the road and run off ditches that were used to keep the road from flooding. They have bad rains here once in a while.

There were barbed wire fences that looked like they stretched for miles. The posts looked old and the barbed wire was rusty, like it had been there forever. Vines and brush entangled the fencing.

There was an old sign but I could not read it, years of being in the weather had taken its toll.

The grass that covered the sides between the ditches and the black top was about ankle high. Looking to the right, it looked about the same; less the old sign.

I had been living in the city for a while so the next logical step for me was to grab my cell phone and make a call to Charlie for assistance.

I'll get to Charlie in a bit, at that moment I realized I had been away from the country to long because there was no cell service what so ever and I kind of got upset for a minute,

JUST LIKE I WOULD OF IF I WERE STILL IN THE CITY.

I ALSO SAW THAT THERE WERE NO VEHICLES TRAVELING ON THE ROAD EITHER, JUST ME AND MY BROKE DOWN CAR. THEN IT HIT ME; I WAS EITHER GOING TO HAVE TO WALK, OR SIT THERE AND WAIT FOR SOMEONE TO STOP. I FIGURED I MIGHT AS WELL GO FOR A WALK.

STEPPING BACK THROUGH THE RED SLIMY DIRT I MADE MY WAY TO THE CAR, OPENED THE DOOR, SAT DOWN IN THE SEAT AND REACHED OVER TO THE GLOVE BOX. THERE WAS SOME PLANE WHITE PAPER IN IT; I NEEDED TO WRITE A NOTE, CAR BROKE DOWN BE BACK AS SOON AS POSSIBLE, OWNER.

REACHING BACK INTO THE BACK SEAT I PICKED UP TWO WATER BOTTLES, TURNED AROUND AND TOOK THE KEYS FROM THE IGNITION SWITCH, STEPPED OUT OF THE CAR AND LOCKED THE DOORS.

I THEN PUT THE NOTE UNDER THE WINDSHIELD WIPERS ON THE DRIVERS' SIDE, TURNED AND WALKED BACK THROUGH THE SLIME AND ONTO THE BLACK TOP.

Stopping for a few minutes to think, my first thought was; will my things in the car be okay? Realizing there was not anything of value in there anyways, I was in the middle of nowhere; or so I thought.

The next question occurred to me of which way to go, back the way I came from or just keep moving forward.

The last town had been about ten miles or so back. Out in this part of the country towns are spaced about twenty miles apart. About a days' ride on a horse or so I was told by someone who claims to know these types of things. So; Forward I decided to go.

By the way, the reason for my journey is simple enough. I am going to see an old friend that I have known for twenty plus years.

He had called me about two weeks ago. We were talking about some things that he had been through lately.

He had mentioned a couple things about some people he had just met a few months back, what they were trying to accomplish in the small town they lived in and how I might find it interesting.

He didn't go into much detail. Bottom line, I think, he just wanted to get together and have a few more good times.

It had been a while since we'd seen each other. I needed a break from the projects I was working on anyway so I said to him "see you in a few weeks". I could tell in the sound of his voice that this made Charlie very happy.

He's a happy kind of person anyway, country at heart, a good ol' boy and fun to be around; always making light of a situation that might not have a promising outlook.

He does have a serious side and it is not one that you want to be on. Overall: a crazy fun loving person. A little old school mixed with new.

He gave me his address, said it was a small town in Missouri. Well; compared to where I had been living.

Anyways; back on the black top, I started down the road to Charlie's.

Along the way I was thinking about random things, looking at the country-side. As I walked I noticed that there were wild flowers and the trees were very tall and dark green.

There were birds of all colors flying around calling to each other, wide open pastures with little creeks running through them. Even the old barbed wire fences looked really neat the way that someone had repaired them in places with old tree branches and wire.

The antifreeze smell from my broken down car was gone. I could smell the trees, grass,

FLOWERS, ECT. IT WAS REFRESHING TO MY NOSE AND MY MIND.

IT WAS AT THAT MOMENT I HAD TO LAUGH. WHAT I DID, PEOPLE DO ALL THE TIME IN THE CITY BUT IN THE COUNTRY NOT SO MUCH BECAUSE OF THE OBVIOUS; I HAD WRITTEN A NOTE SAYING, "CAR BROKE DOWN", WITH MY CELL PHONE NUMBER.

YEAH; LIKE ANY ONE WOULD LEAVE A PERFECTLY GOOD RUNNING CAR ON THE SIDE OF THE ROAD AND GO FOR A WALK AND WITH OUT OF STATE PLATES, IN-BETWEEN TOWNS TO BOOT WITH MY CELL PHONE NUMBER, THAT CAN'T BE USED IN THESE PARTS.

THERE WERE NO CELL PHONE TOWERS. AFTER I HAD A GOOD LAUGH AT MYSELF, I STARTED BACK WALKING COUNTING THE FENCE POSTS AS I WENT DOWN THE ROAD. I CONTEMPLATED THE THOUGHT OF;

What Would Charlie Do!!

Before I give you my thoughts on what Charlie might do; let me tell you a little about Charlie. Maybe you can figure it out before I write it in this volume.

Part Two:

Charlie..

So Charlie's a few years younger than me; raised on his grandfather's farm in Missouri during the sixties and seventies. It was a big farm, little over five hundred acres.

The nearest town was about fifteen miles away. Small town population in those years was about twelve hundred. The kind you may see on T.V. these days if you don't get out much, but they were real back then.

I know because I also was a part of that culture at one point in my life. It was a real educational experience, one I will share with you in the pages of this book.

Charlie lived with his grandfather, grandmother, father, mother, two brothers and his sister that he rarely talks about. His

UNCLES, AUNTS AND COUSINS ALSO LIVED NEARBY.

Generations of his family lived in this area of Missouri. They were well recognized. The big family was an important key to the success of the farm. It was largely populated.

Hogs about two hundred plus and cattle about the same. Chickens, horses, rabbits also had a fish pond that was about six or seven acres. It took a lot of work and most of the family worked on it.

Charlie's family grew a large amount of corn and other types of feed for the livestock. They grew a large garden too, but their main source of income was logging.

They were known to be the best in the area and they took pride in that status. They logged on their farm and surrounding private and commercial corporation properties. They cut an astronomical amount of board feet in that area.

Charlie rarely talked about his youth, from the times between his birth and about eleven years old. I really didn't ask him about it.

He talked about fishing trips to the rivers, creeks and lakes in his area; how beautiful the landscape was around these places and the incredible fishing.

They would bring back ice chests full of fish, cat fish, and all kinds of pan fish. They would let some of them loose in their pond then deep fry and eat the rest.

On these outings a large majority of his family would go. Barbequing, swimming and playing ball were all part of the day.

Taking hunting trips and Friday night football was another thing they did together as a family. There were gatherings put on by their customers usually to say thanks for a job well done. Like I said they were well known and respected loggers.

Around the farm the list of chores he was responsible for was not overwhelming. He spent a lot of his time riding with his grandfather in the trucks and on the equipment; watching and learning how they worked and what it took to fix them and keep them running.

Once I asked him about T.V., what he might have watched. Thinking maybe we had watched the same shows; Gilligan's Island and Star Trek; things he might have liked about the shows.

He just laughed and replied "Didn't watch much T.V. It was for watching the news after diner. Might have watched a little if you were sick lying on the couch, didn't get but two or three channels anyways, mostly listened to the radio".

He did talk a little about his cousins and when they were around. They would build forts, ride bikes and hunt with their BB guns, the usual things farm kids do, PLAY. They

JUST TRIED TO ENTERTAIN EACH OTHER.

The majority of stories he shared with me started about the age of fourteen or fifteen when he was going to this small school, getting up early to catch the bus. It was about a thirty minute ride.

The old school was built like an H, two long buildings two stories high with a covered walk way in the middle to keep you dry while you changed classes during the rainy or snow seasons. He knew everyone; they all knew him and his family.

Normally you would think this was a good thing, it would give you respect or acknowledgement but Charlie didn't quite get that though. He was a little odd compared to the rest of the kids his age.

Being raised around older men and listening to their stories, along with being there for his grandfathers and fathers business deals; Charlie became aware of money and the effect of it in business and in life.

In business he learned its over head, fuel, labor, wear and tear on the equipment and the effect on profit. Even at an early age he learned how to repair equipment with very little, sometimes making parts.

He learned these things by spending a lot of time with the men around the farm and watching them succeed and fail at their tasks. This made his mind-set and vocabulary different from the other kids he went to school with.

When it came to school; they were studying history, math, reading and writing. At the tenth grade level Charlie was thinking about how to fix some old corn binder without spending a large sum of money. Keeping the cost low would put some money in his pocket.

Charlie told me he had already learned

how to read, write and count money. He also knew he wanted to be a logger and work the farm like the rest of the men in his family.

He told me about a party he had gone to while in high school. Fifteen or twenty kids all went down to some field where they all hung out. Some of the kids had bought beer and a few bottles of wine with the money they got from their parents, allowances or whatever.

Charlie had a few things to do on the farm before he could go. So he showed up a couple hours after dark. They had had a big fire burning, they were all drinking and their car radios where on.

At this point in Charlie's life he didn't really drink but he had a couple of beers while he was talking to a few of the boys he knew from logging on their parents properties.

Later, they ran out of alcohol and no one had any more money. Charlie had some but he didn't say anything. Instead he asked them how they got the money to buy what they had and why they didn't have some kind of job to support their habits; that you needed to work hard if you're going to party hard.

Taking offence to what Charlie had said, some slope head started talking about how Charlie dressed for school. The fact of the matter is that he liked what he wore.

He liked the cowboy look, the blue jeans, long sleeved shirts with a t-shirt underneath, his favorite boots he wore everywhere and his cowboy hat. Plus that's just the way the men in his family dressed.

About the time everyone started walking to where all this was happening Charlie walks over to a pile of firewood, picks up a big branch, swung it hitting the kid right in his ribs and curling him up right in front of everyone. They all froze.

Charlie just turned and walked towards the sled he drove up in, jumped in and drove away, never saying a word. It's really no wonder why they all thought he was odd; he is but in a good way.

Charlie did say that one good thing came out of that party.

The next Monday when he had gone to school a girl came up to him and started talking to him about the party and what he may be doing the next weekend. Charlie said "she must have liked the way he took care of business. Must of excited her if you know what I mean"; end quote.

They started hanging out, going to the movies, driving around, swimming, and going to parties in the fields and woods. There wasn't really much to do in the small town.

During this time he had explained to her how he had felt about other boys and the way that they acted around her and the way that she reacted to them. In short he was the jealous type.

She was a good looking girl who liked to talk so it was normal for her to have girls and boys around her talking about whatever the gossip was that week.

Charlie had not been raised around that kind of outgoing, in the middle of everything kind of female. She tried to reassure him that there was nothing other than talking going on. Well I think we all know what s going to happen.

She was doing what she does; it was in the morning about ten thirty on a Thursday when it all came to a head. The bell rings and it's time to change classes. Charlie gets up out of his old wooden school desk and walks to the door and enters the hallway.

Upon entering the narrow hallway Charlie could see all the class doors where open and that it was lined with gray lockers. The teachers were monitoring the kids as they moved from class to class.

On the way down the hallway he could see this boy with his arm around his girlfriend, talking to her with a big smile on his face. Charlie didn't know at that time, but what happened next would change his life forever.

Apparently this guy did not hear the story of how Charlie was known to take care of business. Charlie says he walked straight up to the kid and punched him right on the side of his head knocking him to his knees. The kid grabbed his head with his hands.

Then Charlie kicked him in the stomach with his boot, by this time a crowd of kids gathered. They were shouting, Hit him Again, Fight Back and so on.

Just then someone grabbed Charlie's left arm and tried to break up the fight when Charlie turned around and swung his right fist, hitting them. That person let go of his arm and Charlie turned his attention back to the kid he was inflicting punishment to.

Two other boys had jumped in the middle of them and were protecting the kid.

So Charlie turned again back to who had grabbed his left arm, it was his math teach and that right fist had landed right on the side of his face.

Two other teachers then grabbed him. Charlie knew that that one moment of being out of control of his own emotions cost him a trip to the office.

So the teachers escorted him back down the hallway. They passed the reception area and went straight to the principal's office where they set him down in a chair directly in front of the desk.

The principal was not there. They told him to sit there until he showed up and then left the room.

Charlie's mind began to wonder, he looked around the office. There were trophies and books on the shelves, pictures on the wall with plaques and posters that read, Just Say No to Drugs and Be All You Can Be.

The little radio was on in the corner, you could just barely hear it. He started starring out the window and could see the pupils passing between the buildings on their way to class.

He began to think, how bad could this be?

In his own defense, he didn't intentionally hit the teacher.

Maybe they would suspend him for three or four days or make him work on the school grounds for a period of time.

He then heard on the radio that it was going to be a beautiful day and to make the best of it. At that moment he realized the grave situation he was in and what the consequences might be.

Entering the room, the principal along with the teachers asked Charlie what had happened. Charlie told his story the best he could.

They walked out of the room to the reception area. He could see them talking, their facial expressions and their hand movement through the glass in the door that separated the rooms but he could not hear what they were saying. Charlie knew it wasn't good.

They returned to the room that he was in and asked him if he remembered what he was shouting during the fight. He replied "Not really, I was mad and just went off".

The principal told Charlie that he needed to call his parents and have them come up to the school. At this point he knew it was serious. So he picks up the phone to call the farm house.

Part Three:

The Farm House

He dials the phone number on a rotary phone. This is the early seventies. Cell phones are not wide spread yet, it's rare if you see one at all in these parts of the country. It's a land-line only to the farm.

Some of you may remember when land lines where hung off telephone poles that were made from trees, soaked in creosote and shaped like crosses. Believe me there were times when I thought I knew why they were shaped that way.

Before that if something happened, emergency, or not, it was all about you. If you were able, you might go to your neighbors' house but if you lived in the

COUNTRY BACKWOODS OR ON A MOUNTAIN TOP THEN THAT OPTION MAY NOT BE POSSIBLE BECAUSE OF DISTANCE, TERRAIN, WEATHER, ECT.

You had to know some first aide. Time is crucial; it can be the difference between life and death. Most people in these areas, young or old, knew basic first aid.

They were taught at school or just by watching their people perform it when necessary. You too should know some first aid, its important.

Speaking of neighbors, before cell phones if you lived in the country you better have had a good relationship with them because if your car was broke down or you needed a little help with something, odds were they were your closest help.

So if you were pissed off at them or whatever, you learn to work your difference out QUICK. It's just about respecting each other.

Odds are if you are broke down on the side of the road it was going to be your neighbor who drives through and believe me you want them to stop. It may be a long way to town or an hour before someone else shows up. So be nice to your neighbor.

By the way I did not mention smoke signals because that was before my time but trust me as I tell you this story there was a smoke signal involved once in Charlie's life.

Well that day I bet Charlie wished he didn't have one of those land-lines because he knew if someone did hear the phone ringing it wouldn't be his father. He would be working somewhere on the farm or out logging.

The reason I said if someone heard the phone is because there was only one of them in the old farm house. Charlie described to me as, well it was big, six rooms and a bathroom.

The bathroom had an old cast iron tub

that was huge. It had those legs on it that bowed out. There weren't no shower and the toilet they had to flush with a five gallon bucket sometimes.

Well that was mostly in the winter. It had a sink, mirrors, shelves and a wooden rack they hung things to dry on.

The other six rooms consisted of a kitchen that had a real nice wood stove, a big wooden table his grandfather built for his grandmother; it was about twenty years old and beautiful. The chairs were high back and covered with leather. They had designs carved on them too.

It also had a refrigerator, big double sink and pots hanging on the walls. There was a lot of counter space, cupboards and a lot of cooking stuff his grandmother and mother were always cooking with.

Another room is where they sat after dinner. They would watch a little T.V. if there was something on but mostly listened to the radio, talk and things like that. It had a couch, chairs and some tables.

The other four rooms were bedrooms. That's all he ever said about the bedrooms. I guess he thought that's all I needed to know about the bedrooms. He did say it had a porch that wrapped just about all the way around the house and a tin roof.

The phone was in the room where the T.V. and radio were. The kitchen and bathroom were on the opposite side of the house. That's where his mom, grandmother or anyone else would be. Nobody would be in the sitting room during the day.

The men would be out working so Charlie finished dialing the phone and puts the receiver to his ear and it begins to ring, then ring and ring.

He tells me that he began to think that maybe no one would hear it. There weren't

No answering machine. He would hang up and go from there. The principal will just have to send him home after school.

He would of just had his mother or father come to the school the following day to talk to him, that way it would give him time to get his story straight and let things cool down a little.

The phone rang five or six more times. Back then it would have rang for hours if you let it but on the next ring he hears a hello. It was his mother. He begins explaining the predicament he's in, tells her his father needed to come to the school.

His mother explains that he was thirty or forty minutes away and asks to speak with the principal. Charlie hands the phone to the principal and she explains to him as well.

He tells her that someone was to be there to sign Charlie out. She states that it would have to be her and that his father could come down the following day to find out exactly what had happened. The principal agrees, tells her thank you and hangs the phone up.

The principal tells Charlie that his mother is on her way. Charlie silently sighed in relief. He begins to think that this will give him more time for things to cool off a little.

Sitting there his head filled with thoughts on the situation at hand, you know, the good, the bad and all the outcomes, trying to imagine if anything good could come from this.

Time fly's by and the next thing Charlie hears is the door opening. He looks up to see his father. Immediately Charlie begins to feel sick. His hands begin to sweat not knowing if his father was going to blow up on him.

His father just looked at him with one of those I'll be with you in a minute looks, turns to the principal, puts out his hand for a shake and says sorry for the boys behavior.

They walk into the principal's office leaving Charlie outside of the room while they talked. Charlie's mind really began to run wild. He's not in there to defend himself; like there is any real defense; What Happened, Happened.

Some time passed and his father walks out with a stack of papers in his hand. Looking at Charlie he says, "Let's Go"!

Walking through the school on the way to the parking lot he could see the facial expressions on everyone as they passed. Word had spread fast about what happened.

They reach the truck and the only thing that was heard, was the slamming of the doors the cranking of the engine and then the radio playing. They listened to the radio all the way home. Nothing was said.

Arriving at the farm, shutting the truck off his father told him to go to his room and

THINK ABOUT THE DAYS' EVENTS AND AT DINNER THEY WOULD DISCUSS THEM.

AFTER A WHILE HIS MOTHER KNOCKED ON THE DOOR AND SAID THAT IT WAS TIME FOR DINNER. THERE WAS NO WAY CHARLIE WAS GOING TO BE ABLE TO EAT BUT HE WAS THIRSTY. HE ALSO KNEW THAT IT WAS TIME TO FIND OUT HIS PUNISHMENT AND WHAT WAS GOING TO HAPPEN WITH HIS SCHOOLING.

THEY WERE ALL SITTING THERE AND HIS FATHER BEGINS TELLING CHARLIE WHAT THE SCHOOLS DECISION WAS. HE WAS TO BE EXPELLED FOR THE REST OF THE YEAR, WHICH WAS ONLY ABOUT FOUR OR FIVE MONTHS AND HE MAY BE ABLE TO RETURN THE NEXT YEAR TO THAT SAME SCHOOL OR ANOTHER; EITHER WAY HE WOULD HAVE TO TAKE THE TENTH GRADE OVER.

THEY ALSO DISCUSSED HIS ATTITUDE, MAINLY THE WAY HE DEALT WITH HIS EMOTIONS. BOTTOM LINE, UNTIL THE NEXT SCHOOL YEAR HE WOULD BE WORKING ON THE FARM AND IN THE FAMILY LOGGING BUSINESS; WHICH MEANT GETTING UP with his father and working with him until the day was done, six or seven days a week, whatever it took with little or no pay and no lip.

They finished eating and he was excused from the table. Charlie washed up and went to bed. Thinking he had gotten off pretty easy considering the facts and he eventually fell asleep.

The next thing he remembered was his fathers' voice saying "get out of bed Charlie", looking around at the clock; it was four forty in the morning. His father then told him to meet him in the kitchen.

They ate breakfast not saying a word, standing up and pushing his chair under the table his father looked at him and said "It's time for you to learn what it is to work for a living". He had meant it too.

Charlie's father didn't get a real good education in school. He had worked on the family farm his entire life with his father through some hard times. There wasn't anything he couldn't do on the farm. They had built the farm up to what it is.

Charlie's father would tell him, "If you're not going to be smart and get a good education by going to school, reading books and studying to get a good knowledge of how to work with your brains and a pencil, then you better learn to be physically strong. Without a good education there's nothing but labor in your future and that requires a strong back, arms and legs.

You need a lot of common sense about what you are doing; if you don't have these things your future doesn't look to good.

The only way to get these things is to get up early six or seven days a week and go to work, learning these endeavors of your choice.

Becoming the best takes time; mistakes will

cost you both in money and physical pain, so you'd better learn to be tough; persistence is a must.

They left the table and walked to the truck carrying their lunch, stopping to fill a five gallon cooler of water. There's no convenient store in the woods. You carry in what you need for the day.

Charlie jumps in the truck, slamming the door behind him and boldly saying "Ready to go". There was a moment of silence.

Then his dad said "What are you doing? We don't start anything on this farm without checking the oil and water first. You're working for me now".

Charlie realizes it was going to be a long day. He had gone to work with his father several times before but he could tell in his father's tone that it would be different from now on.

He wasn't worried, he knew the work.

Complying with his father's demands he re-entered the truck. Starting it, down the road they headed to the area his father had been logging for the last four or five days.

His father explained to him what they were doing on the parcel of land they were working on but not much else was said. This made Charlie feel a little odd and couldn't figure out why his father wasn't talking to him about the problem at school.

Arriving on site where they would be working, they were greeted by his uncle and the crew. Charlie knew them all. Charlie's father told his uncle to put him to work and that he would be back in a while, he was going to check on another job.

His father had just made it out of hearing range when they all started in on him. It all pertained to the mistake he made at school. The language they used to describe his behavior was colorful and vulgar to say the least. This went on all day while they were

LOGGING.

Despite all the raucous talk, the day went by smoothly. Nothing had broken down. It was a dry clear day so no equipment got stuck in the mud and a large amount of wood had been cut safely.

It's a blessing when days go by smoothly in the woods logging or on the farm tending to the livestock or equipment. It's normal for things to break down. The equipment is older and takes a lot of abuse.

Weather, hours running and terrain are all a factor. Oh and the livestock; if you've ever owned a pet well then times it by a hundred or so you might start to get the picture and don't forget the crops and gardens.

Easy life, it's not. Especially when things get hectic, it takes a certain personality to endure the long hours, tough labor and few days off. They lived by the old ways, Work

Hard - Play Harder.

Remember now, there really wasn't much to do in small towns like these but whatever they were doing, alcohol was usually involved and turned these times into crazy adventures.

Some of these escapades seemed too far out to be true but after knowing Charlie for a year or so I began to realize that they were true. I could relate to some because I also have had quite a few stories that seem unbelievable.

We've lived similar events in our lives which I believe brought us together like brothers. We trusted each other. These kinds of events build a hardworking, wild, crazy, honest and loving soul that has very little fear of the "what if's" along the way. You learn that you have to think it, be it and then just do it.

Having faith becomes a real turning point

and you have to have it to live this type of life. Without faith I do not know how people like me and Charlie could get up out of bed day after day trying to move forward, sometimes just barely moving, but looking for our place to be physically and spiritually.

Before I go any further let me tell a few stories he told me so later when I write about Charlie and I, and the things we did, it won't seem like I'm telling lies or made up stories because we really did do some crazy things.

Some of them were dangerous; some just funny but these are what made us good friends. Some of our adventures even stopped us from going insane; some of them we just did for fun. It is who we are and how we were raised.

As I had mentioned earlier, Charlie's dad worked on the farm his whole life. He learned through trial and error how to keep the farm going and its equipment. If it was broke Charlie's Father and Grandfather would tell him just fix it like they had done.

On the farm there was a big barn, several pole barns and a bone yard. They never threw anything away. It was all recycled.

You didn't really think that recycling was something new now did you. Depending on what it was, determined where it went; barn, pole barn or bone yard. These places were full from years of collecting.

If something was broke, go find the part. Odds were, it was somewhere there. You might have to take it off from something in the bone yard or take two, disassemble them and take the good parts to make one. It didn't matter what it was, a starter, alternator, drive shaft, body parts, bolts or even a shovel and rake.

You were taught at a young age. First it was stand there and watch, then it was hands on. You learned to work on everything using the welders, grinders, cutting torches, and all types of the tools. Some of them had blades and some of the others had chains.

On the farm you had to know something about being a mechanic, machinist, veterinarian, electrician, roofer and a plumber.

A jack of all trades and you had better be good at them all because break downs and sickness would set you back on completing your chores. The longer it took to get them done the less time you had for relaxation. You learn to fix things the right way the first time.

Charlie had never been in this position before but with the circumstances at school it was time for him to learn some old school ways, the type of schooling that would be helpful to him the rest of his life.

His teachers would be his grandfather and

FATHER. THEY HAVE ALL THE "CREDENTIALS AND EXPERIENCE" WITH FRIENDS AND NEIGHBORS WHO COULD VOUCH FOR THEM.

THEY HAD BEEN AROUND THESE MEN A LONG TIME AND WATCHED THEM ADAPT TO THEIR CHANGING TIMES, YOU KNOW, WHEN LOGS WERE AT THE LOWEST PRICES IN YEARS, BEEF PRICES WHERE LOW AND UPKEEP WAS MORE, OR THE WEATHER SEEMED TO BE TO DRY OR TO WET.

THESE MEN ADAPTED TO WHATEVER CIRCUMSTANCES WERE GIVEN TO THEM. THEY MIGHT NOT HAVE ALWAYS HAD THE RIGHT EQUIPMENT OR TOOLS BUT THEY IMPROVISED AND GOT THE JOB DONE.

THEY HAD THE BRUTE STRENGTH WHEN IT WAS NEEDED BUT THEY ALSO WERE VERY SMART. THEY COULD THINK FAST, MOTIVATING FIVE OR SIX PEOPLE TO WORK TOGETHER IN A MOMENT'S NOTICE; THEIR SKILLS HAD BEEN DEVELOPED OVER MANY YEARS.

THEY WERE RESPECTED BY THEIR COMMUNITY FOR

deeds they had performed in the past without being asked, they knew it needed to be done so they just stepped up and took care of it.

They had their flaws but they kept a positive attitude the majority of the time. When they were on their game, though you could feel their ora, they could make the task at hand seem easier. Things just went smoother when you worked for them.

Around these men you learned to be persistent because "can't do" meant "can't stay". You had to be the "can do it" man. Hours working on any given day meant nothing.

The old saying "whatever it takes" meant just that to them. So it was, get your head in the game and do whatever it took to get the job done.

I think at this point I will just tell you one of the stories that Charlie told me. It happened about three or four weeks after he was expelled from school. It was one of his first major learning yet thrilling experiences.

He said it was raining that day and that there was no way anyone was going to be able to get into the woods to work. It would have been way too muddy;

So Charlie, his dad and a couple of other men decided they would spend the day working on equipment in the barn.

It was the big barn on their property. Two stories high and it had large rolling doors that would keep the rain off from them while they worked.

They could also keep warm if it became too cold because the first floor had an old round wood heater. They also burned old used motor oil in it.

His grandfather put a drip system on it years ago. They would burn anything in it if it came down to it though.

A lot of the tools were kept on the first floor as well. Jacks, jack stands, torches, the welders, compressors and parts that were needed for them. The upper floor had everything to fix what you were working on that the lower floor didn't.

Anyway, they were going to be working on an old truck, a tractor and a log hauling trailer. Charlie's dad put him and a friend of his uncles working on the old truck. It needed the brakes done and the exhaust welded back up.

His grandfather would be changing the oil and greasing the tractor; just giving it a once over; nothing major.

His dad and the other man would be working on the log trailer; it needed a lot of grinding and welding done. It was a major job that had to be done right.

These old log haulers took a lot of abuse from bouncing around the woods, having a ton of logs dropped on them, crossing the creaks and sliding into trees and boulders on the way out of the woods. They still had to be able to travel down the blacktop safely to town, or face big fines from D.O.T.

So in the area Charlie and; well, we will call him Roy because Charlie forgot his name; They were jacking the truck up so they could weld the muffler. They just needed it raised a little bit, the tires wouldn't even leave the concrete.

They would roll under it and with the acetylene and oxygen tanks, use some old metal clothes hangers and a few pieces of scrap metal that they had cut off an old barrel, patch and hang the muffler.

It would be welded and made good as new or better than any muffler shop up town could do and at very minimal cost. It was something they did all the time to save on the overhead.

They finished the muffler and moved to the front of the truck to put the brakes on. They all ready had the brake shoes, keeping a spare set around was the norm back then because all you had to do was take an old pair down and have them relined with new material.

Being prepared was a must; Town was not all that close and the parts store weren't always open. No twenty four hours a day, seven days a week stores in these small towns.

People had lives, families, places to go and things to do. Work was important but not their driving force.

So while Charlie went up stairs to get the set of brake shoes that they would be using Roy started jacking up the truck with an old bumper jack he'd taken out of the back of the truck instead of walking across the barn.

The barn was about five cars wide by the way. You could move around the equipment

THAT THEY WERE WORKING ON THAT DAY WITH PLENTY OF ROOM; THEY WERE IN THE FIRST BAY, THAT'S WHERE ALL THE TOOL BOXES WERE. THE JACKS AND JACK STANDS ALONG WITH THE OTHER TYPES OF BULK TOOLS WERE IN THE LAST BAY.

ROY, FOR SOME REASON DID NOT GO GET THE FLOOR JACK OR THE JACK STANDS. HE WAS CHANGING ONE SIDE AT A TIME. HE FINISHED THE PASSENGER SIDE WITH NO PROBLEM, LET THE JACK DOWN AND MOVED TO THE DRIVER'S SIDE OF THE TRUCK.

HE THEN TOOK OFF THE WHEEL AND DRUM, AND REPLACED THE BRAKE SHOES; NO PROBLEM BUT AS HE WAS CHANGING THE BRAKE SHOES HE HAD NOTICED SOME WIRES HANGING DOWN BY THE FRONT OF THE TRANSMISSION.

SO ROY WALKED ACROSS THE BARN TO THE LAST BAY WHERE THE CREEPER WAS, PICKED IT UP, WALKED BY THE FLOOR JACK AND JACK STANDS, TWICE; AND WALKED ALL THE WAY BACK ACROSS THE BARN TO THE TRUCK.

He put the creeper on the floor and rolled under the truck, found out what the problem was and figured out what tools he needed and was on his way back out from under the truck; that's when the incident happened.

To make it easier on him he grabbed the frame of the truck so that he could wheel the creeper out faster. When he did the truck shifted under the pressure. The old jack slid out from under the bumper and the truck fell on Roy.

He was pinned under it, still lying on the creeper. He couldn't move. He yelled "HELP" but it wasn't very loud after all he was pinned.

Charlie was at one end of the tool boxes and heard him. He turned, saw what happened and yelled for his dad to come over. "Roy's pinned under the truck".

Charlie, his father and his uncle lifted the

side of the truck enough that Roy could get out from underneath of it. Roy was very lucky this day. He was not seriously injured. He didn't even go to the hospital but the next morning Roy was very sore. He figured he might have cracked a few ribs. It could have been a lot worse.

What saved Roy that day was that he put the tire back on the old truck, if he had not put the tire back on the truck the total weight of it would have been on him because the clearance would have been minimal.

Charlie said that he learned something very important that day. What he had learned was that safety comes first. Not letting yourself be caught in a position where you could be pinned or smashed.

Roy walked by the floor jacks and stands twice and just didn't take the extra three minutes it would of took to do the job in a safe manner. From that day forward he always tried to think of safety first when he was working, especially when he was working crews of men.

Once you see something like that happen; it tattoos an impression of that experience in your brain so every time you start working it pops up. You will with no doubt find a way to prevent the incident from happening again.

So, if you don't have jack stands or side bars you will find a way to improvise and make it safe as possible. If your brain is working at all you don't want that "I KNEW BETTER" statement to come out of your mouth because QUIT follows next.

Back to the second part of our story;

After we found out that Roy was not seriously hurt, they decided to have a few beers. Kind of a celebration for Roy's good luck, Charlie was allowed to have a few also.

While the men were drinking they were

telling stories of their past. Charlie really enjoyed these stories. He was laughing with the men, thinking about nothing but what was happening right then. He felt good.

One thing that really intrigued him was how often the men would mention God. How if God hadn't been with them that night they would have never made it out of that bad situation or God was their co-pilot that night or Thank God for someone helping them out of that jam, ect, ect.

As a child Charlie was never around the men when they had been drinking, telling these types of stories about close calls, the trouble they had been in or the successes they had experienced.

It kind of made him think that somehow he was becoming closer to the men in his life and as time went on he was allowed to hear more of the stories. He was also allowed to have a few beers during these times.

They were stories of good times and bad, it made Charlie want some of his own to tell. He liked the stories best when they were about their accomplishments. Which made them stand out in the group of men!

How they took nothing and made something, how they were able to get out of a jam successfully and even how they would buy something for five dollars and sell it for twenty five. Charlie always talked about the men.

Where he comes from the males were outside most of the time taking care of the labor part of the farm. The females were taking care of the paper work for the business and the family needs. They worked as hard as the men. That's just how he was raised.

The men around him would say that their women were the smart ones; that they could think things through when it came to things such as paying the bills and paper work.

It was a very important part of running the farm and the business. With them taking care of it, the men could concentrate on the hard labor of the farm and business such as the equipment, animals, and crops. The women did take care of the gardens and animals as well.

There would be times when the women and men would do it together with the children. These would be the fun times together. They may go fishing for the weekend, they would swim, fish, eat and play. They would sit around a large campfire talking about their lives as a family.

The men would tell stories of their wives and children. They would also talk about what was happening on the Farm, in their community, church and business.

By the way the men in the community did go to church, just not that often. The women would talk about what was going on with other families in the community as well but

It was considered gossip when they did it.

Charlie would say "I think they called it gossip just to get a reaction from the women". But overall it was never the things that the men talked about when they were just having a few.

As time went on Charlie saw that the fishing trips, swimming holes and the gatherings they attended seemed the same. The men never spoke of the crazy things they did; their accomplishments or the close calls they all talked about when he was only with them. He wondered why.

So one day Charlie was riding to work with his grandfather in one of the truck's. He waited until they left the blacktop.

He said the reason why he waited to get to the dirt road to ask him is that the old truck was so loud at 60 mph that you couldn't think.

The truck had mud grip tires and the roar

Going down the road was crazy; not only that but the truck shook and the doors rattled, but he did say it was the friendliest truck in town because the fenders waived at everyone; it was a little rusty.

Anyway, it was about a twenty minute ride through the woods to the site where they were logging. You had to go slow because the ruts, rocks and branches.

So as they were riding Charlie just came out and asked his Grandfather why. His Grandfather just looked at him with a big smile; paused a moment Then said,

"Ya' know Son; some of those stories may be a little questionable. They do have some truth to them but after having a few, the stories get spun a little differently".

Another words they may enhance them a bit to make them funnier or more dramatic. Not all the time, but if you hear them enough times you will hear them change.

"You're still young enough that you've only heard them once or twice. They are exciting and full of drama. The women have heard these stories for years and have also heard them spun many different ways.

Odds are, they might have been there for some of these things that happened or they know the true story of what happened and it may not be so funny or exciting to them.

The drama is just that to them and the women just don't want to hear it. It becomes an annoyance to them; Especially after a few".

Charlie's Grandfather chuckled a little and then continued.

"The small children don't need to hear them at all because there'll be plenty of time for that later down the road. Do you understand what I mean by that?"

Charlie said "Yes" and his Grandfather added.

"But us guys after having a few tell them because we don't care how many times the stories get told, we get a kick out of them. Heck some of those stories we were all together, yet we all tell them a little different".

Letting out his breath and taking in another he says,

"There is something special in them that bonds us together as we share them over and over, it brings back a special time in our lives even though they may change just a bit. We also have our own individual stories that we tell and they change as well if you listen to them enough times".

Charlie takes a second glance at his Grandfather and nods his head acknowledging his words,

"Plus Charlie, you know that we swear in those stories and say things you just shouldn't say around women out of respect.

That's why your dad tells you to watch your mouth when you're around women, it's just different when it's just the guys, it's why you were taught to say yes ma'am or no ma'am, it's why if you're mother needs help you just do it.

It's also why you open the door for them, letting them walk through first and there are a lot of other examples but the bottom line Charlie, is that they need to be treated different".

"They keep us going with all their hard work, from the paper work for our business, keeping the home up and running, not to mention all the errands that need to be run. Plus Charlie, They put up with us and all our MAN ways"!

"Keeping a good woman is a hard thing to do; It takes a lot of work living out here in the country. So keeping momma happy is very important to all of us.

The old saying goes, If Momma Ain't Happy, Nobody Is. So when you see a boy get, should I say, brought to attention when he acts out around a group of women it is probably because he wasn't talking right or he was disrespecting them in some manner".

His Grandfather continued: "Us men like to call it an attitude adjustment if you will, enough to get him thinking on the right tract. So respect the women in your life, they will be happy and so will you Charlie".

They pulled up to the job site where everybody was working. Charlie smiled at his Grandfather, gave him a nod of the head and they both opened the doors, stepped out and went to work.

Like I had mentioned earlier; Charlie really wanted some stories to tell of his own so about three months of working in the woods was all it took. His chance had arrived. This is how he tells it.

They left the farm one morning, it was hot that day. His uncle and him were to move the equipment to the new jobsite about thirty or forty minutes away from the old site they had been working.

They pulled up to the old site in an old semi with a lowboy trailer attached. They loaded the log skidder on the lowboy, chained it down and started loading up an old pick-up truck that they left in the woods to move tools around with.

There were chains, wedges, gas cans, bars, ropes and a lot of little things that needed to be loaded. His uncle helped him for about ten or fifteen minutes then told Charlie he was going to head out.

It would take him sometime to get the skidder through the woods and in the mean time Charlie could load the old truck up and about the time he would get the skidder out of the woods Charlie would have the pick-up loaded and would be heading out. That

WOULD PUT THEM THERE ABOUT THE SAME TIME OR GIVE OR TAKE TWENTY MINUTES OR SO.

Charlie said okay and kept loading the pick-up. He finally had it all loaded and sat down for a minute, drank a bottle of water and started the old truck up.

He was about half way through the woods when he noticed the old truck was overheating, He could smell it, he could also hear it boiling in the radiator.

Charlie stopped the truck and the steam rolled out from under the hood. Opening the hood he looked around to see if a radiator hose had broke or fell off.

Glancing at the radiator to see if he had knocked a hole in it or something had stuck in it he noticed the fan belt was not on the alternator.

At that point he knew what the problem was because on this old truck the fan belt only went around the crankshaft, water pump and alternator. There wasn't any power steering or air conditioner, one belt worked everything.

Sometimes the fan belt would slip off and fall on to the crank. If it became loose all you had to do was put it back on and tighten the alternator up and away you could go.

This time Charlie could not find the belt, it must of fallen off or broke somewhere back on the dirt road.

Walking back to look for it would be a waste of time, so he looked behind the seat of the truck for a spare belt. That's where these old loggers kept them and they usually had one there. Not this time, no such luck.

He looked in the big tool box but no luck there either. So Charlie sat down in the front seat for a minute to think about what he should do.

His uncle was already gone, it'd be a hour or more before he started thinking about

WHERE HE WAS AND IT WOULD TAKE ANOTHER THIRTY MINUTES TO BACK TRACT AND FIND HIM. SO WHILE CHARLIE WAS SITTING THERE HE BEGAN TO THINK ABOUT MAKING A BELT, JUST ONE GOOD ENOUGH TO GET HIM OUT OF THE WOODS.

THE OLD TRUCK HAD A SIX CYLINDER IN IT WHICH MEANT THAT IT DID NOT NEED A LOT OF POWER. IT WAS PRETTY SIMPLE REALLY, JUST GET UP AND DO IT. NOTHING LOST, HE DID HAVE TIME TO KILL BEFORE THEY CAME LOOKING FOR HIM ANYWAY.

SO HE BEGAN DIGGING AROUND FOR A PIECE OF ROPE IN THE BIG TOOL BOX. IN THE BACK HE HAD FOUND SOME ROPE, IT WAS NYLON AND HE KNEW IT WOULD NOT WORK BECAUSE IT WOULD SLIP ON THE PULLEY'S. IT WAS SLICK LIKE PLASTIC. WHAT HE NEEDED WAS SOME COTTON OR TWINE ROPE AROUND A HALF INCH IN DIAMETER.

LOOKING IN THE BED OF THE TRUCK HE FOUND WHAT HE NEEDED. IT WAS COTTON. THAT WOULD WORK HE THOUGHT. SO HE PUT IT AROUND THE PULLEY'S FOR MEASUREMENT.

He had already loosened the alternator all the way. That was the only adjustment that had to be made. So he cut the rope and began to braid the rope ends together.

After braiding the rope he took some bailing wire and tightened both braid ends. With a little help from some duck tape he covered the wire. He placed it around the pulley's.

To Charlie it looked possible, so he tightened the alternator up just enough to turn. He then took a five gallon jug of water out of the bed of the truck, filling the radiator up; he cranked up the motor.

When he walked back around to the front of the truck to see if it was working, he said he began to laugh. It was working just fine; the water pump and fan were turning. He finished filling up the radiator and put the cap back on.

Shutting the hood he told me that he just

couldn't stop smiling, jumping back into the truck he slammed the door, put it in gear and slowly started out of the woods.

The temperature gauge didn't work but the volt meter did. He could watch that. As long as the alternator was turning the volt meter would show thirteen or fourteen volts. That would let Charlie know the water pump was turning too.

He was also smelling the air vents and listening. If there was a problem he would be ready to shut the truck off before any harm could come to the engine. Nothing happened he made it all the way out of the woods and safely back to the blacktop.

He stopped there for a minute and thought, could he go down the blacktop? The truck had never been over twenty or thirty miles per hour in the woods.

So he figured that he would ease it down the road keeping the R.P.Ms low not getting over forty miles per hour, forty five at the most. So down the blacktop he went.

He was about three quarters the way there when his dad flashed the headlights at him. He knew that meant to pull over. He did.

His dad walked over to the old truck and Charlie stepped out. His dad asked him why he was going so slow. He began telling his dad the story. His dad laughed and told Charlie he had to see.

Charlie opened the hood of the truck. His dad looked at the make-shift belt. He put his hand on Charlie's shoulder and said "Great Job. That's what we call whatever it takes to get the job done around here"

His dad shut the hood and told Charlie to go ahead to the job site, that he'd already went this far and he was pretty sure that the truck would make it the rest of the way and that he would follow him the rest of the way there.

Later that evening, all the men were having a beer out by the barn. Charlie was having a beer with them and as usual they were telling their stories. Charlie didn't say anything.

So Charlie's dad looked over to him and asked; "Hey Charlie, You got any stories you want to tell?" Everyone stopped and looked at Charlie.

Charlie told me at that moment he felt something he'd never felt before; pride, excitement, joy; He didn't know what to call it but it was his first story. His Only Story, and now he was going to share it with the men in his life.

He told his story. The men were all laughing and patting him on the back; telling him there was no way he pulled it off and he was full of it. They carried on and on. Charlie said he was on top of the world.

He finally had a story to tell and he would

TELL IT OVER AND OVER. CHARLIE CARRIED THE BELT HE MADE THAT DAY FOR YEARS IN ALL OF THE CARS AND TRUCKS HE OWNED; ALWAYS READY TO SHOW IT.

❖

WHILE I HAVE BEEN WRITING I HAVE CONTEMPLATED ON IF I SHOULD WRITE MORE STORIES OF CHARLIE'S AND HOW MANY? HE HAS ACQUIRED MANY. THEY COULD FILL A BOOK AND THAT'S JUST THE ONES THAT HE HAS TOLD ME UP UNTIL NOW. HE ALWAYS COMES UP WITH MORE.

I'VE HEARD ONES ABOUT PEOPLE WHO HAVE LOST THEIR LIVES IN ACCIDENTS, FUNNY ONES AND SERIOUS ONES WHERE HE REALLY GETS INTO TROUBLE AND ONES THAT WERE KNOWLEDGEABLE GIFTS THAT HE CAN USE THE REST OF HIS LIFE.

THEY ALL HAVE BEEN ACCUMULATED OVER A LIFE TIME. SOME, TO TELL YOU THE TRUTH SOUND FAKE BUT ARE NOT. I ALSO THINK THAT IT IS IMPORTANT TO GET WHERE HE HAS TO MAKE A DECISION ABOUT school and the things that happened to him from that point of his life.

The stories are no longer child like or just living to make them happen. The young teenager has to make decisions that set him on a path; a path that some will look down on.

Others, who have already traveled the path, will respect him for just surviving some days. His path will take him places both good and bad.

I have decided That I would like to write you one more story, one that I can relate to: a similar situation happened to me while I lived in Alabama.

❖

Charlie and a Friend wanted to go to town, it was a Friday night. Everyone would be where the kids hang out. They would need something to drive.

Charlie knew that out behind the barn was an old 68 Chevelle. It was a two door. It ran, looked cool and the AM radio worked. It was a six cylinder automatic transmission car.

The guy's parked it because it had one tire that the cords were showing though, the other tire was flat and the front A-Frame bushings were completely out of it.

They looked around the barn for some tires to put on it but came up empty handed. Earlier they had bought a six pack of beer to take to town with them and have a few with their friends.

Well they opened a beer sitting there trying to figure out what to do. Charlie remembered that there was an old Lemans sitting out back with some other old cars.

His dad was a General Motors' man. When

it came to cars, people back then were like that. It wasn't unusual for people to stick to one brand for their entire life especially on a farm. Parts would interchange.

The Lemans had a bad motor but the tires were still fair. They had a few miles left on them. He would have to wait until his dad came home before he could do anything with the Lemans. He needed permission to take the tires off it.

In the meantime Charlie and his friend jacked up the old Chevelle, took the tires and rims off and put the car on jack stands so it would be ready if his dad said they could use the Lemans for a parts car.

They cleaned the old Chevelle up, checked the oil, water and transmission fluid all while listening to music. They had also opened another beer. Everything was going good. They were having a good time.

About a hour and a half had passed when

His dad arrived. They waited a few minutes to let his dad settle in. Then Charlie walked up to the house and asked his dad about the Lemans.

His dad said yes to go ahead. He also said that he didn't know if they would interchange with each other, one had drum brakes all the way round and the other one had disc brakes up front.

Sometimes they wouldn't fit right but the tires would fit, they were all Fourteen's.

So they jacked up the Lemans, took the tires and rims off and went back to put them on the Chevelle. They fit right on, no problem or so they thought.

They tightened up the lug nuts, jacked the car down, washed their hands, jumped in and cranked the engine.

They looked at each other smiling like the cat that just ate the bird; Radio on Charlie's friend said "We need to get another six-pack".

Charlie said "Yeah" and dropped the car into drive and stepped on the gas.

The car wouldn't move. It was like the brakes were on. Charlie checked to see if the emergency brake was on, it wasn't.

He put the car in park and got out to see if they were hung up on something, it was all clear

Getting back into the car he pushed the gas pedal to the floor. When he did, the rear tires spun. The front didn't move but the car slid forward about a foot.

Turning the car off they got out and looked at each other like what's going on. Puzzled by the event his friend handed him a beer. It was the last one.

They set there a minute then Charlie remembered what his dad had told him about the rims.

They jacked the car back up and tried to spin the tire. It wouldn't move. They looked around to try to figure out the problem.

They found it. The rim was too deep. It was up against the A-Frame, stopping it from moving. So they decided to finish their beer and think on it for a minute.

They came up with the idea to put the tire on backwards. It would stick out from the fenders some. They also didn't fit quite right against the drum.

By now they had a buzz going on from the beers they had drank. They were also out. So they tightened the lugs as tight as they could get them without snapping the studs off.

They spun the tire to make sure it wasn't locked up. It spun freely. They did the same to the other side, backed the car out and headed for the store.

Charlie said to me that it drove terrible. As

he told me this I thought to myself No Kidding. You Really didn't need to tell me that.

He continued telling me that they had to stop a couple of times to tighten the lugs nuts but they made it to the store where they bought another six pack. They didn't try riding around though.

They sat there for a while, had a couple more beers and talked to people they knew who were also getting beer from the little store.

The people they knew would just shake their heads when they looked at the car. They had a few laughs, finished their beer and headed back to the farm stopping a few times to tighten the lug nuts.

The next morning his father seen the car and asked Charlie about the ride. Charlie told his dad the story. His father just laughed and asked him if he'd learned anything.

Charlie just smiled telling his dad people think "I'm crazy, but I did learn something".

Never put the car on the ground until you spin the tire to make sure it spins free and not locked up. You also won't go far with it mounted backwards, but it can be done.

I too had a situation where I had to put the tire on backwards but it was on a trailer. I was out in the backwoods, it was my only option. I had to keep checking the lug nuts to make sure they were tight, but I did make it home.

Back then there were no cell phones and you had to be creative! The thing about being raised in the country, where the population was small, is that there was not any traffic.

You were able to try these types of things. The only thing that could happen is that you tear up a car, truck or maybe your tractor. You learn the limits of what can be done

WITH OLD JUNK CARS.

MOST OF THE TIME YOU WERE NOT ABLE TO BUILD ANY SPEED, PLUS THE LOCALS WEREN'T APT TO HARASS YOU IF YOU WEREN'T SPEEDING OR DOING ANYTHING THAT WAS GOING TO HARM OTHERS.

MOST PEOPLE WOULDN'T TAKE THIS TYPE OF STUFF UP TOWN ANYWAY. TOWN WAS FOR BUSINESS NOT FOR MESSING AROUND.

IF YOU WERE OUT IN THE COUNTRY AWAY FROM TOWN, ODDS ARE YOU'RE NOT GOING TO SEE POLICE. THE SHERRIFF WOULD HAVE TO BE CALLED AND IT WAS RARE TO SEE A SHERIFF.

PEOPLE USUALLY TOOK CARE OF THEIR OWN PROBLEMS. THEY WERE NEIGHBORS; THEY HAD TO SETTLE THESE THINGS. NO PLACE TO HIDE OUT THERE, JUST THE WOODS, YA'LL AND THE CRITTERS.

SO THE END OF SUMMER WAS ABOUT A WEEK AWAY; THE TIME TO MAKE A DECISION ABOUT SCHOOL WAS NOW. THEY NEEDED TO GO TO THE SCHOOL AND TALK TO THE PRINCIPAL TO DETERMINE WHAT WAS TO HAPPEN.

They made an appointment with the school secretary. She told Charlie's father it would be a couple of days because the teacher and principal both had to be there at the meeting.

His father replied "Okay" and told her to call the house to let him know the time and ended the call with "Thank You Ma'am".

Turning to Charlie he said

"You know, that teacher did not deserve what happened to him. In life you need to be able to control your emotions, at least just a little bit. In your life there will be people who hurt you physically and emotionally.

Using your fists will not always solve the problem and it could cause more harm to you yourself and others involved".

"I want you to think about what happened so you can talk to that teacher, make him understand that you are truly sorry for the pain you caused him and that the terms they set for your return to school you will comply to and take as punishment for your outburst".

His father paused for a moment. The next words from his mouth, Charlie said, had a major impact on his life.

"Charlie; Being smart isn't enough"!

"You really have to care about what is happening; Thinking about things takes time. So when you are upset, puzzled or confused about things, take a moment.

Some things may take days. You never know.

You might have to pray on things; Ask for his guidance; Listen for his answers; Look for it. Believe and Know that it will be reveled to you".

This caught Charlie completely off guard, he'd never heard his father talk this way, the tone of his voice and his father's facial expressions.

Charlie's father ended the conversation with

"You're not always going to find someone to believe in so you better know what I believe and who I believe in.

You have a couple of days to think about what you are going to say to that teacher, I won't just take sorry for the answer and you're going to explain yourself to this person".

He left Charlie standing there in shock, walked to the barn and started working on one of his projects.

Charlie said that he had lost time and didn't remember how long he'd stood there but eventually he went to the barn and started helping his dad.

They worked for a couple of days when the call came in. The time was set, it would be that afternoon. A lot of thoughts had passed through Charlie's head.

After the conversation with his father he figured he would go in the office and just tell them those thoughts and maybe they would make some sense out of them.

The day had come and Charlie and his father walked in together. They had the conversation; Charlie didn't ever say what was said, just, that he was supposed to start the tenth grade over. There was also some voluntary work he would have to do.

When Charlie said the word Voluntary to me I laughed.

They left the building and went back to work. He had until that coming Monday. That is when he would have to start back.

During that time he did a lot of thinking while they worked. The thoughts of what would the other kids at school think, would he be an outcast and would they consider him a "bad boy" so to speak of.

He knew of no one else that had been in the

same kind of situation that he was in.

That soon pasted because he really didn't care about them that much. They weren't family. Family was the Important thing. He had told me that it also crossed his mind if he really wanted to go back to school at all.

He had been working on the farm with the men as a man. He was paid the same as one of the new men that worked logging. He enjoyed the work.

What he was learning would help him be a better asset for the family business. He was learning fast and was told that by his father and that made him feel good about himself.

He knew his father wanted him to go to school though. Education was very important to his father because he had learned the hard way.

He had told Charlie several times that being book smart wasn't the only thing he needed but it was very important, that it would make his life easier when dealing with others in business and life in general.

To Charlie working with the other men during the time he was out of school was like nothing he had experienced before.

Time went by fast and it wasn't a struggle keeping focused. Charlie said bottom line he was afraid at this point in his life he would miss something because it was all new and exciting.

Other things also came into his mind if he didn't go he would miss out on the school events, would he be invited to parties and what would his girlfriend think. Sunday, Charlie thought all day on those kind of things.

Time pasted very fast until that evening. They were sitting at the table when Charlie's father asked him what he thought about school.

Charlie had talked a little bit to his father

about his thoughts. His father knew he couldn't make Charlie learn at school, sending him there would be a waste of time for the teachers and Charlie if he didn't want to go.

He replied "I will go and try".

Charlie's mother told him thank you. His father said he thought that he was making the right decision. They finished eating their dinner and Charlie got up and went to bed.

The next morning Charlie went to school, checked in with the principal and then went to class. The day passed by with no problems and at the end of the day Charlie worked around the school grounds for about a hour or so cleaning up.

The days passed, then weeks. Some days he would mow the lawns or weed eat and trim out the school yard.

After two months or so the principal called Charlie into the office. The

Conversation was short; he told Charlie that his punishment was over but not to do anything that would get him back in trouble.

There would be no second chance. Charlie answered with a prompt, yes sir. The principal told him to return to class.

A few weeks later Charlie started drifting off in class. His mind just wasn't there at school. He was thinking about his father and the men on the farm, what they were doing, logging, working on trucks or cars or maybe even tending to the live stock.

Charlie said that his money was also disappearing. He wasn't making any either and that bothered him.

His teacher called on him a few times to answer a question, realizing Charlie wasn't paying attention, he told Charlie that he wanted to talk to him after class.

He stayed after and talked to the teacher, he told the teacher the truth about what was going on in his mind. The teacher responded and told Charlie that he shouldn't be thinking those kinds of things and that he should be studying and paying more attention to what was happening in class.

The response of the teacher was not what Charlie had expected. He thought the teacher would have more insight on the matter.

The short statement made him upset with the teacher and his thoughts questioned how this person could be so non-caring, after all he had told the truth.

The more Charlie thought about it the madder he became. How could his teacher show such disregard for his emotions!

He started to tell the teacher where he thought he should go but at that moment something told him just to be quiet and go home after school and talk things through with his father.

Charlie told me this was the first time he remembers just taking a minute or two to think about things.

When Charlie saw his father later that day he told him what had happened. He and his father talked for a couple hours about how to take care of the things that were bothering him.

At the end of their conversation the outcome was that Charlie and his father would go to the school and take him out. Rules were that he could return if he hadn't been expelled.

In the meantime Charlie would prepare himself to take the G.E.D. test while working on the farm full time. And so that is what they did.

Time went by, Charlie passed his test, made money working and bought himself a car. Nothing fancy but he fixed it up and everyone knew that it was his car. It was fast for what it was and people around town though it was cool.

A lot of the people around town had a good attitude towards Charlie; they saw that he was really trying to put his life together. They would invite him to parties and outings.

His girlfriend was still by his side and everything was moving just fine in his mind. The work became harder but was no problem for him.

It went this way for about two years then one night out on a date with his girlfriend she told him that she thought that they should try spending time with other people, that they were young and before they became too serious or she got pregnant she wanted to see and experience different things.

He said she told him some other things but he could never recall what they were.

At that point he had lost it, not outwards but inside. This was the girl he thought he'd marry. His first thoughts were too serious; he was already there.

Secondly he thought who did she want to see besides him and what was it that they couldn't experience as a couple, they were always together.

He thought they were having fun, he thought she really loved him. Charlie was in shock and caught off guard.

Charlie kept his thoughts and didn't say anything to her right then. He pulled the car over and sat there for a bit. Finally he looked at her and said "okay".

He told me, looking back, he knew it was over the moment he looked at her and said okay. Her face had changed in a way that he couldn't describe. Nothing was said after that.

He turned the car around and took her

straight home. She got out, walked to the front door and went in. He drove off and headed straight to the beer store, went in, picked up two six packs and headed to the barn.

There he sat drinking, listening to the radio and thinking about what had happened. He was hurt like nothing he had ever felt. He tried to think of ways he could make her love him the way he had her.

A lot of things went through Charlie's mind that night but it was getting late. So Charlie just went to the house and found his way to the bed.

He knew he had to work the next morning and letting his dad down was not an option for him.

The next day everyone knew that something was wrong but Charlie said nothing. He just kept working. The weekend had come, it was Friday night and Charlie decided to go over to some friends house. They too all knew what was going on.

They jumped in someone's car and went for a ride to the drive in movies in the next city over. There were people there they all knew and they were all having a good time. Charlie was not saying much though.

There were other girls there. Some even liked Charlie but he couldn't get his mind off her. They were all drinking, mostly beer.

When the movies were over they all went home, seemed like everything was okay.

It went like this for about a month or so when he said the anger and the hurt took over. He began drinking a lot on his days off. The boys that he'd been hanging out with were all into it.

Charlie had the money to party hard. He was still working everyday at this point. He had his car and knew people. The boys he was hanging out with all knew a lot of people

TOO.

THEY HAD PLENTY TO DO. GAS WAS CHEAP AND SO WAS THE LOCAL HAMBURGER SHOP, THIS IS WHERE CHARLIE ENDED UP EATING MOST OF THE TIME.

I AM NOT GOING TO WRITE ABOUT THE THINGS HE DID DURING THIS TIME; MAYBE LATER.

I WILL TELL YOU SOME OF THE STORIES HE TOLD ME ARE LIVING PROOF, OR AT LEAST TO ME, THAT THERE IS A GOD.

HE SHOULD HAVE BEEN KILLED DOING SOME OF THE THINGS HE DID DURING THIS TIME. HE REALLY DIDN'T TALK THAT MUCH ABOUT HIS FEELINGS THOUGH.

HE DID SAY HIS FATHER TRIED TO TALK TO HIM, AND THAT SOME THINGS HE WAS TOLD MADE SENSE.

CHARLIE WAS CRUSHED INSIDE. HE JUST COULDN'T GET OVER THE HURT. THE THINGS THAT HE AND HIS GIRLFRIEND HAD TALKED ABOUT KEPT going over and over in his head. Emotionally he just couldn't deal with it.

Eventually things became worse, showing up to work so hung over he couldn't perform the work, wrecked his car and had been thrown in jail. Nobody around town wanted to be around him. It was bad.

One day his grandfather stopped him. "I need to talk to you Charlie":

This is how the talk went;

"You're really making a mess out of your life. I know why but that s something you're going to have to deal with. Our family has been here a long time and we are getting tired of hearing about the stupid things you are doing.

I can't even go to the post office without hearing about you.

I have lived my entire life here and made a good name for our family through hard

WORK AND DOING THE RIGHT THING.

YOU ARE MAKING THE FAMILY LOOK BAD AND I DON'T LIKE THAT. DON'T YOU REMEMBER ME TELLING YOU A LONG TIME AGO NOT TO CRAP WHERE YOU LAY YOUR HEAD"!

CHARLIE ANSWERED "YES" AND HIS GRANDFATHER CONTINUED.

"IF YOU KEEP GOING THE WAY YOU ARE YOU'RE GOING TO BE DOING SOME SERIOUS JAIL TIME OR END UP DEAD. EITHER WAY IT'S TIME FOR YOU TO TAKE A TRIP AND GET OUT OF TOWN BEFORE PEOPLE JUST CONSIDER YOU A PIECE OF TRASH.

IF THAT HAPPENS YOU WILL NEVER BE ABLE TO LIVE HERE AND BE A PART OF THE COMMUNITY, LEAVE AND GO TO YOUR UNCLES IN ARIZONA.

YOU KNOW THAT WILL BE THE BEST THING FOR YOU WHILE YOU STILL HAVE SOME MONEY AND YOUR CAR.

IT WILL BE DIFFERENT OUT THERE, THE LAND AND THE PEOPLE, EVERYTHING, AND IF YOU NEED A little more money I will loan you some.

Your uncle will help you find a job, I have already talked to him and he knows what is going on."

"I Love You, you're my flesh and blood but you need to pull yourself together. We both know that it will not happen here though". And then asked;

"Do you understand?"

Charlie answered "Yes".

His grandfathers final words that day were;

"Think about it and then you can come and talk to me".

He then turned, walked away and went back to working.

Charlie went out that night with the boys. They stopped and picked up some beer then went riding around as usual.

This time his girlfriend wasn't on his mind. What his grandfather had said was. It upset him to think he was giving the family a bad name.

What had bothered Charlie the most was the effect he was having on his grandfather. It had never crossed his mind that he might be hurting his grandparents.

He knew how hard they had worked for the respect they were shown by the community.

He went home early that night. Everyone he was hanging out with asked him what was wrong. He just said that he didn't feel that good.

The next day he laid in bed for most of the day thinking about what he was going to do. When he did get out of bed he went to town and ate a hamburger.

On his way back to the farm is when he made his decision. He would go but he needed to talk to his dad about what his

GRANDFATHER HAD SAID AND TO SEE WHAT HE THOUGHT.

He went home, found his father in the barn and asked if he had a few minutes. His father replied "sure" and Charlie began telling him about the conversation that he had with his grandfather.

After talking with his dad, they come to the conclusion that his grandfather was probably right.

His father asked him if he needed anything before he went and that he was a phone call away if he needed to talk.

Charlie asked if he would get the men together to wish him well on his journey, have a few beers and tell some stories of their adventures.

Thinking maybe that would give him strength, confidence and faith to take the challenge that was in front of him. The next morning he would head out to Arizona.

His father told him sure; they would get together Friday night. That would give Charlie two days to get his car ready for the trip. It really didn't need much anyways. He also needed time to pack his bags and say his goodbyes to his friends.

It was settled. The only thing Charlie needed to do right then was tell his grandfather what his decision was.

Then somehow tell his mother and grandmother but he believed that his father and grandfather would help him with that. He found his grandfather down at the fish pond. He began to tell him about his decision to go.

About half way through the conversation Charlie noticed that his grandfather began to tear up. This affected Charlie in a major way.

Instantly he knew his decision was the right one. He could see the relief and the

PRIDE IN HIS GRANDFATHERS FACE.

They talked for a while, His grandfather told him that he loved him and Charlie replied "enough to help explain this to the women of the family?"

They laughed. Charlie knew his grandfather loved him. Charlie never did tell me what was said to the females of the family but that was not that unusual. Conversations with the women were kept private. That's just the way he was raised.

So Friday came and most of the people that Charlie knew showed up to say their goodbyes and best of lucks.

They drank a lot of beer and had a barn fire, a big one. The barbeque was awesome, everyone enjoyed themselves and there were a lot of stories told. The majority of them were positive things like good fortune or the family.

This gave Charlie a sense of pride. He also

drew strength from their stories. The kind that only comes when you know it's the right thing you're doing.

Some family members had gone through hard trying times and made it. Confidently Charlie could say that he would too. There was also his faith, something he knew was real.

The party ended and everyone left. It was just him and his father standing by the fire. His dad began speaking and Charlie's ears perked up and he began to pay close attention.

"Son, I love You".

"What's happening to you also affects me. I hurt because I will miss seeing you. You are me, just a little wilder. The hard times are just as important as the good. It takes both to make a rich soul and that's what is important".

"Tomorrow you will start your journey to

find you. It will take the rest of your life, you know deep down what's right; Try to follow that. If you really get in a situation where you think there is no way out, I will be here.

Work hard and try to save some money, you never know what may show up on your door step.

If you have some money saved you may be able to turn one thousand into five thousand in a matter of days.

You already know those things, there simple. The hard thing for you is how to control your emotions. They are what got you to this point.

They will cause you a lot of grief if you don't learn. There has to be some kind of balance. Your uncle is a good guy at heart; you'll get along with him just fine.

Call me once in a while, Good or Bad and don't let the girls drive you crazy. Go fishing

EVERY NOW AND THEN". LIFE IS SHORT, DO WHAT IT IS THAT MAKES YOU HAPPY WITHIN REASON. COMMON SENSE GOES A LONG WAY."

CHARLIE'S FATHER WALKED OVER AND GAVE HIM A HUG. CHARLIE HUGGED HIM BACK. HIS FATHER LET GO AND WALKED TO THE HOUSE.

CHARLIE WATCHED HIM FADE INTO THE DARK. HE STAYED AT THE FIRE FOR ABOUT A HOUR MORE, DRINKING A FEW BEERS AND LOOKING AND LISTENING TO THE DARK.

IT WAS A BEAUTY THAT HE HAD TAKEN FOR GRANTED. TONIGHT THOUGH HE WAS AWARE OF EVERYTHING AND FOR GOOD REASON; TONIGHT WOULD BE HIS LAST FOR A LONG TIME.

HE COULD SEE THE STARS BRIGHTEN UP AS THE FIRE DIED DOWN AND THE SHADOWS OF THE TREES. THERE WERE MANY TYPES OF NOISES.

HE SAID HE HAD THOUGHT HOW IT WAS A STRANGE MOMENT. HE DIDN'T KNOW WHY UNTIL MANY YEARS LATER. IT'S PRETTY SIMPLE THOUGH.

The fire went out so Charlie walked up to the house, washed his hands and face, then went to bed.

When Charlie woke up he could hear his family in the kitchen. He dressed and went out there and had breakfast with them.

Not much was said, everything had been said the night before that needed to be. He stood up, told them that he loved them all and they all followed him outside.

The car was already packed. They all hugged and prayed for protection. Charlie sat in his car, cranked it up, put it in gear and told them he loved em all once again. He let off the brake and headed to Arizona.

Let me ask; Have you figured out what it is that Charlie would do? If not than just keep reading, I will write it out later but not now.

PART FOUR:

BACK ON THE ROAD

Back on the road I had been walking maybe three miles when I noticed a creek. Being raised around water and loving to fish I had to go look and see if I could see anything swimming around. This however did not turn out like I had planned at all.

As I became close to the creek, I mean right at the point I could look; I must have stepped wrong on a rock or my ankle just gave out.

(It runs in the family, you could be walking on a flat surface and the next thing you know you're on the ground, hereditary issues like these are a wonderful thing)

I fell and landed in the creek. Catching myself by putting my hands out in front of me I was up to my elbows in the water.

Anything that had been swimming in the creek was gone for cover.

I stood up right away like someone might have been watching. My shirt was wet and my paints had dirt on the knees but I seemed to be alright other than that.

I really remember looking up and down the creek as I stood up. I wanted to see a fish because in my mind I knew if I could see a fish it meant the creek was alive and not just some run off ditch.

I didn't walk the creek, I had had enough, falling in the middle of nowhere was not good.

Returning to the blacktop I began singing old Motown music; Before you think that I've lost my mind, I like country music, But the funk brothers and Motown sound good anywhere. That's just my personal opinion.

The road had some curves in it. Just walking around a curve, well enough to see

Ahead, it looked like a building but that was all I could make out.

Getting closer I could tell that it was a little store. Beside it was a little shelter of some type or maybe just something to get out of the rain.

This I thought was good because I had not walked as far as I had thought I might have to. Then the other thought; Are they open?

Just for a split second I thought the "Oh Ship" thought, but it passed when I saw a person open the door. It looked like they were sweeping the floor out.

As I became closer I could see a Y in the road. It sloped off to the Northeast. Then I began to hear dogs I thought. I don't care for dogs that are running lose. Getting bit by one is not on my list. I had no choice I had to keep moving forward towards the store.

Getting closer to the Y I could see an old garage set back a little. The trees came up even with the building on my side of the road. The closer I became I could see more.

There were cars that surrounded the entire garage. Two bays a little office up front with four or five chairs and a desk that looked old, I mean forty or fifty years old.

The whole office would have fit right into the fifties or early sixties if you know what I mean. If you don't, go rent some old car movies and you'll see one. Anyways, the bays were average width but the doors dint role up. It had the type that slide to the side.

These were great doors, easy to open. the material to make them was steal and wood. They used plenty to ensure the quality and workmanship, time and material. You could open them with one hand.

The building could fall apart but not the doors. Looking inside I noticed it had a pit. Ha, when's the last time you have seen a pit?

Some people have no idea what a pit is. If

You tried to get a permit now days for a pit the EPA would have a field day with you and for you that have no idea; let me tell you about pits.

To begin with they are giant holes in a garage bay made for you to walk under the vehicle, usually longer than a truck, three feet wide, give or take five inches or so. You drive over it, don't crash in it; If you owned a garage for any length of time it happened.

It's just one of those things. Changing oil in these things was easy as eating apple pie. You were able to see everything underneath the car. Its setbacks were that it was hot in there, fumes hung in it and water and debris had to be cleaned out of it.

So lifts are a good thing. They are in all your local garages. The other bay had an old one. The inside was packed with a lot of stuff, it was a mess;

(It looked like it was a busy one man show

that was fixing everything and just didn't have the time to pick up behind himself. It was, rush to get this done then off right away to the next. He must have just ran out of time.)

By knowing these types of people I can bet you that he knew where everything was. Their computer was their brains. It stayed on when the electricity went out, no problem.

Move something; Let them walk around the garage and they would be able to tell you what you had moved.

Then I realized where the barking was coming from. Off to the side there were two dogs on top of dog boxes that were mounted in the bed of some old truck, roped so they wouldn't jump off.

They looked kind of ruff; you could tell they were hunting dogs. When they saw me they just started wagging their tails and baying. They were some kind of hound dog.

Most hunting dogs of this nature wouldn't bite anyone, they just want to run, chase down what they were trained to hunt and bark their heads off at the same time.

There was someone standing by the driver's door talking to whoever it was in the driver's seat. They were smiling so it looked like they knew each other. They turned and looked at me; I said

"Good morning, How are ya'll today?" (It's a southern thing, I can't kick it.)

They smiled, said "Good, You must be broke down"!

"That's a fact, just down the road a little, three or four miles". I replied.

"What's wrong with it?" he asked.

"Overheated, the serpentine belt broke; I'll need a belt and some water and then I think it'll be okay".

The next question he asked after I finished

telling them what was wrong with my car was one that I did not expect.

"Do you have some tools with you; I most likely have a belt somewhere around here"?

"Yeah, some" I told him, catching me off guard and bringing me back to how real some people can still be. He then tells me;

"Well, You'll have to hang out here or walk over to the store for about thirty minutes; I have to run down to Jim's and jumpstart his old tractor. I'll probably have to shoot it with some ether to get it to run, Jim probably flooded it again".

"By the way, My name is Ron". I smiled and shook his hand and the other guy smiled. Ron walked around, sat in the passenger side, shut the door and waved. The other guy raised his hand a little and they backed out.

Laughing inside I remember thinking to myself that the driver must have been Jim as I turned around and started walking my way

Over to the store.

Thinking; what a good way to start the day; just in the few minutes I spent talking to Ron I knew the car would get fixed, it was just part of his day. Be back in thirty minutes and I knew he had meant it.

It's just something you learn about these types of people. They know everything about what they're doing.

On the way across the lot, I passed some gas pumps and they looked old! The gas price was a dollar fifty four.

Right then I knew they had not been in service for a while. I noticed that there were some old tires stacked up down the side of the building.

Looking both ways before crossing the street, as I was taught, I could see the tailgate of the old truck with the dogs. They were just-a barking away.

Nothing else, everything looked as it was frozen in time or I was the last one standing.

Crossing the road I was walking passed this fenced off area with two gates. The one gate was about half open. As I passed it a man stepped out; Shock ravaged my mind instantly. He could tell I was stunned, and right away he began to apologize.

Gathering my thought back together; Taking two or three minutes to do and I am sure a lot of you know how it is wanting to kill them first before it becomes funny, trying to stop from crying after just getting the ship scared out of you; He said,

"My name is Gary, I was just digging out the aluminum cans and looking around for good stuff that I could use".

It was where they put their trash from the store and where people dropped of stuff. You couldn't tell though, it was one of those wooden fences that you could not see

THROUGH.

I REPLIED AND TOLD GARY THAT IT WAS OKAY. I DON'T THINK IT BOTHERED HIM ONE BIT BECAUSE HE JUST STARTED TALKING TO ME.

"WHAT'S GOING ON WITH YOU TODAY"? HE ASKED SMILING AND I GAVE HIM A SHORT RUNDOWN.

HE TOLD ME, "I COULD PROBABLY HELP IF YOU WANT, THERE ARE SOME BELTS OVER AT MY PLACE, ONE MIGHT FIT AND WATER IS NO PROBLEM".

"AS FAR AS RON; THAT'S NO PROBLEM EITHER, KNOWING HIM AS I DO, IT'LL PROBABLY TAKE HIM A LITTLE LONGER. HE WON'T CARE EITHER WAY, IF I CAN'T HELP YOU HE WILL".

I TOLD HIM THANK YOU AND HE TOLD ME THAT IT WAS NOT A PROBLEM.

AFTER A SECOND OR TWO I ASKED HIM IF HE WANTED A CUP OF COFFEE AND TOLD HIM THAT I NEEDED ONE BEFORE I COULD DO ANYTHING. HE SAID "NO THANKS ALREADY DRANK A POT" AND laughed.

I excused myself and told him that I'd be back in a few. As I turned I heard Gary say, "I'll be right around here".

I started for the front door and noticed that there were advertisements and signs all over the building. I passed a window but couldn't see much.

The building was an A-Frame type about a story and a little more for attic space and the shingling was brown. It wasn't too big but had enough room for two small bays and a little store.

Opening the door I realized just how small it really was, only about sixteen feet across. In the back there was a small counter with a glass front. The register was on top with some other things.

It looked to be only about six feet long and maybe two and a half feet wide. A shelf was off to one side with a microwave, coffee pot, sugar and creamer jars and small packs of ketchups and things like that.

There were shelves that lined the walls. They were in two by one in a half foot sections, about a foot and a half deep.

Having a small cooler certainly limited your choices of meats; Bologna, Hot Dogs, and of course your burrito's, Red or Green cold drinks and a few other items.

In the middle there was a metal rack, some standing shelves about four foot tall and seven feet long that held a few canned goods such as sardines, hash, beans, some camping gear, bread, a variety of potato chips and paper plates.

The lady behind the counter had told me Good Morning.

"How are you?" I replied.

She smiled and said "Alright".

As I walked over to the coffee pot I began

thinking if I really wanted to have a conversation with this women, after all I had things to do. Pouring the coffee into the cup and looking over at her I asked if there was free refills, she just looked at me.

Maybe I should tell her that I was just joking I had thought but instead I walked to the counter, paid her the dollar fifty that I had owed, turned and took the five steps to the door, opened it and heard her say "Thanks".

Replying "You're welcome" I stepped out and turned to head to the little shelter. The structure was four, four by fours, plywood roof with rolled roofing and about sixteen by eight in diameter. Sun off your face and rain off your back kinda size.

Sitting on the homemade picnic table that was placed under it was Gary. He had a beer in his hand, a smile on his face and when he saw me he said "Its five o'clock somewhere".

I QUICKLY LOOKED AT HIM AND REPLIED "AND IT'S TWO FOR ONE".

SITTING DOWN, WE STARTED TALKING THE USUAL; WHERE YOU FROM, WHAT DO YOU DO AND WHAT'S GOING ON. THIS WENT ON FOR ABOUT FORTY MINUTES WHEN GARY ASKED ME IF I WANTED TO GO DOWN TO HIS HOUSE TO SEE IF HE HAD A BELT SEEING HOW RON HADN'T RETURNED YET.

LIKE GARY SAID, IT PROBABLY WAS GOING TO TAKE HIM A LITTLE LONGER TO GET THE TRACTOR TO RUN.

SO I SAID SURE WHEN HE TOLD ME THAT HE ONLY LIVED A BLOCK OR SO AWAY, DOWN THE ROAD AND TO THE RIGHT. WE BEGAN WALKING, HE ASKED WHAT TYPE OF CAR I WAS DRIVING AND I TOLD HIM.

HE SAID THAT HE THOUGHT HE HAD A COUPLE OF BELTS. WE PASSED A FEW HOUSES ON THE WAY BUT THAT'S ALL I CAN REALLY TELL YOU; GARY IS A TALKER, YOU KNOW THE KIND, HE STARTED TELLING ME ALL ABOUT HIMSELF AND HIS FAMILY.

THERE IS NOTHING WRONG WITH THAT BUT SOME

DETAILS COULD HAVE BEEN LEFT OUT.

We reached his driveway and I could see a big shed in the back of his property, I figured that it was where the belts would be.

When we reached the shed the doors were opened and it was full of parts. Gary said he collected parts and sold them to the locals. A few dollars is better than no dollars and I agreed.

There were small parts for mowers, bicycles, motorcycles, cars and other stuff. He walked over to some belts hanging on the wall and picked out a few.

"We'll try these" he said, and that he needed to get the keys to his truck and grab another beer.

This might have scared some people away but I am used to that type of stuff. I have road with many a drunk man. Then Gary asked "You want one"?

"NO" I SAID AND TOLD HIM I JUST NEEDED A COUPLE MORE CUPS OF COFFEE.

"ALRIGHT" HE REPLIED, OPENED THE DOOR TO THE HOUSE AND WALKED IN.

WHILE HE WAS INSIDE I STARTED LOOKING AROUND. HIS OLD TRUCK LOOKED TO BE A CHEVY HALF-TON, REGULAR CAB, FOUR WHEEL DRIVE AND SOMEWHERE IN THE EIGHTIES.

YOU COULD TELL IT HAD BEEN USED. NOT ANY IN TOWN SHINY 4X4 WITH POLISHED TIRES. YOU KNOW THE TYPE THAT NEVER GETS PUT IN FOUR WHEEL DRIVE.

THERE WERE CHAIRS SURROUNDING A TREE WITH CANS ON THE GROUND. THEY LOOKED TO BE COFFEE CANS, TO PUT CIGARETTE BUTTS IN NO DOUBT.

THE HOUSE LOOKED TO BE A THREE BEDROOM. NOTHING FANCY; IT HAD THE LIVED IN LOOK. THERE WAS A CAGE ON THE SIDE OF THE HOUSE.

I WALKED TOWARD IT AND I COULD SEE SOMETHING MOVING IN IT. TAKING A FEW MORE steps I could see it was a coon and a nice one too, just the right size to eat.

They taste delicious, I recommend trying one if you get a chance. About that time I heard the house door slam. "Ready to go" Gary asked.

"Yea" I replied.

We walked over to the truck and jumped in. He started pumping the gas. That told me it had a carburetor on it. It turned over four or five times then fired up and shut off right away.

He pumped it a few more times and hit the key. It started running but this time it kept running. He revved up the engine a little, reached over tuned the radio, sat back up, put the truck in gear and asked; "Which way"?

Oh yeah I forgot to mention; As we were walking towards the truck I had notice he had a cooler with him, red and looked like it

COULD HOLD A TWELVE PACK.

HEARING THE ICE SLOSH AROUND IN IT I FIGURED IT WAS CARRYING ABOUT SEVEN OR EIGHT BEERS IN IT AND THOUGHT HOW INTERESTING IT MIGHT GET.

I TOLD GARY TO HEAD BACK TOWARDS THE STORE. HE PULLED OUT AND RIGHT AWAY WE COULD SEE SOMEONE AT THE GARAGE. WE WEREN'T THAT FAR AWAY. IT WAS RON, HE RETURNED.

GARY DROVE UP AND ASKED RON WHAT WAS GOING ON AND RON REPLIED FOR A FEW MINUTES THEN GARY SAID THAT HE WAS GOING TO TAKE ME UP TO MY CAR. HE TOLD RON THAT HE HAD A FEW OLD BELTS AND THAT HE WAS GOING TO SEE IF ONE OF THEM FIT AND IF NOT THAT WE WOULD BE RIGHT BACK.

RON TOLD US ALRIGHT AND THAT HE'D BE AROUND THERE FOR A WHILE. GARY BACKED UP THEN TURNED AROUND TO THE ROAD, HE REACHED AND TURNED THE RADIO BACK UP THEN STOPPED RIGHT AT THE ROADS EDGE.

HE PUT IT IN PARK LOOKED AT ME AND ASKED,

you ready for one now? "No" I exclaimed, "Need another cup of coffee before I get started".

Gary just laughed, opened the door, stepped out and pulled a beer form the cooler. He then sat back down in the truck and said "Let's get that coffee".

He pulled up to the store, I stepped out; turned and asked him if he needed anything, holding up his beer he told me no.

I went inside and poured another cup of coffee, paid the lady behind the counter and said "Have a nice day". Opening the door I could hear the radio coming out of Gary's truck. It wasn't real loud like a nice beer buzz but it was a three beer level.

Walking to the door I reached out and grabbed the handle when I did I noticed Gary's right hand. He had a cigarette clamped between two fingers; Those two fingers where dark brown.

Now I knew one thing for sure about Gary; he was a heavy smoker. You have seen fingers like that before I am sure. If you haven't, just look around, you'll see them.

Sitting down he turned the radio down a bit and I pointed down the road in the direction of my car. He swung around and off we went. I thought to myself, high-ho-silver-away; another words here we go this could get interesting.

We drove down the road rock'n rolling to the old rock station. They did play some good ones. We were making small talks nothing serious when we spotted the car. Like I said it wasn't that long of a walk. He said to me; "Are you any good at towing"?

I laughed and said, "Yea been doing it as long as I can remember".

"Well lets tow it to my house, that way we can have some shade". "We can also have a few beers and just take our time, ya know".

It sounded good to me so I just said, "let's do it".

He turned the truck around then backed up. He had a rope. It was a good tow rope. This wasn't his hundredth time. It's just something that he does with no thought, second nature type of thing.

We hooked up, agreed on the hand signals, jumped in the vehicles and off to his house we went. Everything went fine. Making it to his house we pulled it under a big tree, you know the one. It was the one with the chairs and cans they use for butts.

I started taking the rope off the vehicles, so I laid down between them just for a few minutes. When I rose back up, Gary had the chairs in a circle and that red ice chest was in the middle.

The truck doors were wide open and the radio was still on about three beers loud. Rolling the rope up, I put it back in the bed of the truck. Gary was just reaching for a tool box in the bed.

Turning I walked to the car, opened the door and popped the hood. Gary had the hood open by the time I got to the front.

Looking down into the engine bay Gary said "The belt is for sure gone". "Wonder if the water pump is bad"?

"Let's give it a quick once over". I said "To make sure nothing is locked up".

Gary took a drink of his beer then lit a cigarette and started touching the brackets, alternator, and other things like the power steering. I was also looking. Everything looked okay.

We both stood up and I said, "We might as well put a belt on it and see what happens".

"Might as well" he replied. He turned to get the belts and at the same time he asked "Ready for a beer"?

Look I really wasn't ready for a beer yet but being how I was raised; If a man is trying to help you and offers you a beer, odds are real high it's going to be a great day. Plus that's what Charlie would do, Have a beer.

So don't be a bump on a log, start living. "Yea might as well" I replied and reached into the tool box to get some wrenches to do the job.

Gary returned with the belts and some beer. He handed me a beer, I sat down and he started trying belts on.

Opening my beer I took a good drink, reached for one of my cigarettes, lit it and took a drag. I turned around took a couple steps, reached up on the tailgate and let it down. It was just the right distance from the car.

At that moment Gary raised up with a belt in his hand, reached for his beer with the other, leaned back against the tailgate, took a drink then said, "Needs to be a little smaller".

He handed the belt to me and lit a cigarette as I searched the pile of belts for one a little smaller. Finding one I sat it between us and we talked for about fifteen minutes drinking our beer.

He threw his can in the truck bed, grabbed the belt laying between us and started to fit it to the motor. Finishing my beer, I asked Gary, "Ready for one?" 10-4 was his reply.

Walking over to get us another beer I heard a car with a strong motor. (You just can tell if you've ever been around fast cars.)

The sound was coming from the direction of the store. Then I heard the driver push it to the floor or at least close to it any ways. It started coming our way, tires screaming.

Then you heard the driver let off it real quick. Looking over the car was a Nova 69' or 70's model coming in the driveway, radio cranking.

Opening the red ice chest I took out two beers and started walking back. Arriving back at the car Gary said "It's Fixed; We just need to crank it and see what's up with it".

Now the Nova and the guy in it has Gary's attention. Pulling up he shuts the car off. He then asked Gary; How's it hangin.

It was all I could do not to bust out laughing, did I take a trip back in time or what, was my first thought. Gary replied, good then the guy in the car starts telling Gary about somebody they both had known and the problem he was having with his truck.

Gary looked at me and says; "Come over here, I want to introduce you to Mick".

Mick says "Hi, what's going on", I tell him about my car and that Gary was helping me get it running right so I could be on my way.

Telling him that I was going to see an old friend who lives another one hundred fifty or so miles away. Mick said, "Gary is good luck like that, Mr. Fix-It and a friend to all", were his exact words.

I didn't hear what the problem was with the truck, just something about the alternator. Gary tells Mick we had about thirty minutes more on my car then we would go down to their friends house and look at the truck.

(We? I thought, What's up but I just kept it in).

They talked for a few minutes then Mick drove off. Gary walked to the front of the car then told me to fire it up and we'll see how she's doing. It started right up. I didn't get it real hot before I shut it off, seeing the temperature gauge rising was all it took.

I knew something was wrong so that's when I pulled over right away. He said "I will go get some water, just let it run".

While Gary was getting the water, I stepped to the front of the car and bent over and gave it the once over, touching the hoses and looking at the pulley's. I didn't tell him I was a mechanic for years. He never asked, just went right for it.

Gary arrived back with the water, opened the radiator cap and poured some water in. It probably only took a gallon.

We watched it circulate for a minute and then he put the cap back on; "You ready for one, we'll need to let it idle for five minutes or so just to make sure that you are ready to go".

"Yeah" I replied and he walked toward the ice chest. I reached in the car and turned the radio on. I had an Anthony Smith CD in.

When Gary walked backup he handed me a beer, looked at the motor and said "Looks Good. What are you listening to?"

I tell him and he says that he's never heard of him, that maybe he'd get to hear some more of him later. Gary then asked me "What do you think is it going to make it"?

I told him sure that it was just like new.

He shut the hood, I just let the car run. The radio was on, Gary Told me to shut the car off but to leave the radio on and that we'd have a couple. I was already on my third beer.

I really didn't want to have any more and figured that I would just finish my beer, give Gary a few dollars for his help, tell him that it was nice to meet him, thanks and maybe get his phone number. You know the old saying, See you down the road somewhere speech.

Gary had other ideas. Well I did not know that then.

We sat back on the tailgate of the truck. He began to tell me the conversation him and Mick had. One of their friends truck would not charge the battery.

They had put a new alternator in but it still wasn't charging and would he come down there and take a look at it.

He said he would after he finished my car. Then Gary asked me what I thought.

"Let's go for a ride, it won't take long. It's about fifteen minutes away and it'll give you some time just to relax: see the scenery.

For some reason I said, "Let's go".

It was probably the three beers; but anyways I shut my car off and walked over toward his truck. He grabbed the ice chest and put it in the back, opened the driver's door and started the truck.

He then looked over at me and said "find something to listen to, I need to go get my test light and volt meter". "Be back in a few". Then off he went.

Turning the radio knob to some old Rock-n-Roll, I sat back in the seat, lit a cigarette and drank my beer. Didn't take him long to return; Before he sat down in the truck he asked me again;

"You ready for one"?

That must be his thing to say. I have heard a lot of different way to ask -you want another beer- and ones just as good as the other. I said yeah and figured one more for the trip.

He sat down, put the truck in gear, turned it around and pointed it to the driveway. Reaching the road he turned and headed for the store.

"We better get a twelve pack to ride on, you need any cigarettes?"

I told him "No, I'm good". He pulls up to the store and shut the truck off.

He said, "Come on, want you to meet someone".

I got out and walked up to the door with him. Gary opens the door, steps in, looks over to the lady behind the counter and says,

"I need some credit today. Need some beer and things. Ya know I'm broke today".

She smiles at him and says "You're always broke".

He introduces me to her. We didn't say much and he went to the cooler, picked out a twelve pack, grabbed a bag of chips, walked back up to the counter and told her to put it on his bill.

She just smiled and said see ya later. We walked out, jumped in the truck, cranked it back up and headed down the road.

After a minute or two Gary said to me, "I have known her a long time".

There was a tone in his voice; one you could tell that there was a story or two between them.

Then he smiled and started talking about the guy we were going to see; that they have known each other for twenty five or thirty years and that he lived on an old farm with an old house that was super cool.

He had a cow, rabbits, chickens and an old horse that they never rode. He told me that it was just part of the scenery.

The guy was a gun nut. He had all types of handguns and rifles. Gary told me that we'd probably shoot a few before we left.

Thinking, this was a strange day but a good one so far, we rode down the road talking and listening to the radio for about ten minutes.

Then we turned on a dirt road. The trees were old, you could just tell. It wasn't that wide of a road; if someone were to come towards you, you'd have to slow down or pull over and stop, just depending where you were at on it.

We go a little bit down the road when Gary pulls over, gets out, unzipped his pants and grabbed two more beers all while taking a pee. Sits back in the truck and says to me, "You got to go?"

I stepped out and took care of it. When I turned around to sit back down I noticed he had put another beer right where I sit. I picked it up and sat it between us.

As we continued down the road we went around some curves. The road wasn't rough, it wasn't real smooth either but not bad. I have been on a lot worse. After two or three miles we came to an old gate and drove through it.

The driveway to the old house was about a quarter mile. There was pasture fenced off on both sides. Going toward the house you could see an old wind mill and the barn.

There was a garden off to one side, not real big maybe fifty feet by fifty feet. I could see the bright red tomatoes and the corn stalks. There were a couple big trees in front of the house where Mick was.

His Nova was parked underneath one of those trees. An old Ford truck was parked in the driveway; If that's what you want to call it.

A box full of rocks, that's what it looked like to me. Really; Two by six's on three sides and rocks sliding out the front;

Anyways, the old truck had the hood up and I figured that it must be the problem child. There was a Blue wade pool, you know the type every big store sells them. One foot deep maybe four foot around.

Well it was under the other tree. Oh yeah, children, four or five of them running around, in and out of the wade pool, two dog's, Mick and someone else.

They were leaning up against the Nova's side panel and the trunk was open. I also noticed that there were bikes and other things related to kids in the front yard.

As we began to pull up beside the truck I figured out real quick why Mick had the trunk opened; Music; At about a six pack level.

I began to admire the house; Two stories, tin roof, lots of windows, two chimneys, and a porch that wrapped around the sides. I could see the paint was that old time white.

If you haven't seen it before I'll try to explain but you should really go and see one for yourself before their all gone.

The white paint on the wood begins to flake just a little over a period of years. The weather, pollen, dew and dirt have set in the cracks and you can see the wood in places, not a lot though, just enough so that when the sun hits it, it can look a little brownish gray.

In the winter it looks different. My

explanation doesn't make sense to me either, it's something you'll have to see.

The steps up to the porch were made of rock and cement. It looked good, not perfected by any means but eye catching and on the porch sat two women.

Gary pull's up, shuts the truck off and hops out. He stretches his arms up just like he got out of bed. Thinking, what's up with that, I step out of the truck and look to my right.

Mick and the other person were almost to the truck. The kids come running up screaming, Uncle Gary, like I wasn't there. Gary's now at the tailgate, he sticks his hand out to the mystery man. The man grabs Gary's hand and Gary yells to me

"This is my life brother, Roger", and at the same time they shake hands, lean in to bump shoulders and as the kids are all hanging on Gary, he tells Roger who I am and says "I told you, Ya never know what you might find at the beer store dumpster".

They look at me and laughed. I just said Hey and that's where we met. Gary asked Roger what was up with the truck. Roger then explains.

We all walk to the front of the seventies Ford truck. There are four of us, just looking; Not much room. Gary's got his test light in his hand, already.

He tells Roger to fire it up and he does. It ran for about thirty seconds when Gary reached over and pulled the positive cable off the battery. The truck dies right away.

"It's not charging" Gary yells, not to loud but everyone could hear, "Crank it back up" and Roger does.

Gary takes his test light and probes the wires on the alternator; I was watching and notice the problem right away, the wire that needed power; had none. Gary noticed it too.

The problem was that someone had messed with the wiring harness; They had the tape off from it. There was no site of a loom. They were all over the place; Wires that is.

"Shut it off", Roger complies, "Just turn the key on, Don't crank it" Gary yells just a little louder than before. Gary starts to probe the wires and I start opening my computer up to the late seventies.

The first thing I thought of was the fusible link. I didn't say anything right away I let him probe for about ten minutes or so then I said a friend of mine had a truck like this and it did the same thing.

Turned out, it was the fusible link on the old truck. You could tell that Gary's brain clicked. He went right to where they were and started probing.

He found it, walked back over to his truck, opened the tool box and took out a piece of wire, some cutters and tape. In front of the

Truck again, he reached in, cut the wire in two places, removing the bad link, added the other wire, twisted them together and taped em'.

He then leaned up, pulled the hood down just enough to see over it, nodded and said "Start it" and Roger did.

Gary pushed the hood up with his right arm, reached over with his left and disconnected the positive battery cable. This time it stayed running;

"It's fixed" Gary said loudly.

Mick said "You're the man" and Roger shut the truck off, walked around front, patted him on the back and said "Thanks brother".

Gary replied right away "I'm not taping this thing up today! Let's show this city boy how we party around here".

Apparently Gary felt comfortable around me; enough so that he wanted to show me a good time. We just met but sometimes things just click with people.

Where Gary came up with the idea that I was a city boy, I have no Idea. Maybe he was just joking; City boy I am not! I will explain later.

He shut the hood and we all walked over to the back of the truck. That's where the ice chest was, Remember "Let's have another" Gary says.

I began to count beers in my head. It was four or five, maybe five or six. Looking up at the sky I figured it to be twelve thirty or close to one and no I didn't have a watch; Who wears them?

Cell phones rule the time of the world but not out here. It didn't matter anyway the sky was blue; The temperature was around seventy to seventy-five with just a very little breeze.

You could smell the flowers, the cows, the

children's horses and the dogs. It was a farm, ya know.

I had a decision to make; Stop drinking or find something to eat and take a nap in my car for a couple of hours before I go. Or I could find out what's on Gary's mind but either way I need to call Charlie and let him know what was going on.

So we're all gathered around the truck talking and just bull shipping about one thing or another when out of nowhere Gary sais to me;

"Looks like you're staying at my house tonight. We just got started and we will find something to eat in a little while".

About that time Roger said; "The old lady 's making Butter Beans, Greens with Ham and Biscuits. We won't starve", then he laughs.

I follow up and say; "Well I need to find a phone, my cell phone has no service out here". They all laughed at that one and Roger explains that here was one in the house. I thought to myself, Oh It's On Now.

Gary and his friends seemed to be good people and I was raised around these types of men, women and kids. Hell, I might as well enjoy the ride. Gary grabbed the ice chest and we all started walking towards the Nova.

I had figured I would wait a hour or so before I called Charlie. That way I can just get the feel of things around here first.

As we became closer to the Nova you could hear the music and it sounded good; Rolling Stones and at a six pack level. We all gathered around the Nova and Mick turned the radio up to a seven beer loud.

The women on the porch started to move around, Mick yells at them to come out to the Nova for a minute.

They yell back and say that it'd be a minute.

Now the kids are running around telling Gary to take them to the store. "Ain't nothing there" he replies".

The tallest one tells Gary, "You always say that". "We want some candy, you know that".

Then I hear Gary say; "We're going up there in a few and I'll bring you some back".

Roger looks at me and asked if I liked fishing. I asked him if a bear ship in the woods. He looked at Gary,

"Let's take him down to the creek; maybe we can catch a catfish or two for dinner".

They agreed and said "Let's go but first we need to take him up on the porch and introduce him to the girls before they think something is going on". They all kinda chuckled and mumbled a few words.

Gary and Roger walked up the rock steps first, I was behind them and Mick was behind me with three or four kids in tow. When I stepped onto the porch the smell was awesome. You could taste the food, it was so thick.

Gary asked the ladies how they were all doing today, they smiled and said that they were fine. Roger introduces me to them; one was his wife, Mindee, the other her sister.

I really do not remember her name. He then asked when dinner was going to be ready. "When its time" she replied.

The girls then head to the door, Roger right behind them. Gary, Mick and I stood out there a minute or two then Roger comes back out; "Let's Go". "She gave me a couple of hours" and we start out towards Mick's car.

Gary stated that we needed some poles, Roger heads towards the barn and we kept walking to the Nova. We arrive there and Gary checks the ice chest.

"We need beer". He picks it up and starts walking to the truck and Mick tells him that he would meet him there and Gary tells me that he won't go anywhere without his car.

At the truck we wait a few minutes drinking a beer. Gary turned the radio to about a four beer level. Roger show's up with a few poles, a five gallon bucket and a little tackle box. He puts it in the bed of the truck and commands, "Let's Go"!

We all jump in the truck, turn it on and head out the driveway. When we reached the blacktop I saw that Mick was right behind us.

"Watch this" Gary said.

He pulls out real slow in the direction of the beer store, going only about forty or forty-five he weaved a little left. Just as Gary pulled it back to our lane you heard Mick floor the Nova.

It came around us tires just a smokin. Gary

and Roger just started laughing. "He will do that every time" Roger said.

"He will be waiting there for us, out of the car, radio on with a beer in his hand" Gary added. I just laughed a little.

It's not that far to the beer store, but sure enough, when we pulled up to the beer store it was just like Gary had said.

We got out and walked to the front of the store. All three of us walked in one after the other. The lady behind the counter said

"Here comes trouble" and then asked if we were all headed to the creek. They all started talking bull with her. I don't remember all what was said but I walked over to the cooler and picked up two twelve packs and headed for the counter.

Gary and Roger were picking out a few things as I put the beer on the counter and asked for a pack of little cigars. She reaches up pulling them down she asks

"Are you sure you want to be hanging out with these boys". "You know they're a little crazy. Nothing bad or anything like that, I mean that they are just crazy".

Smirking I asked, "Should I take that as a warning?"

She smiled and Gary said as he put a bag of candy and a bag of chicken liver on the counter. "The liver's for the catfish and the candy is for the kids later. Ha-Ha"!

Mick had some type of wine that apparently someone had made around here and where he picked it up from I couldn't tell ya. The first time I was in here I certainly didn't see any wine.

He showed the bottle to the lady and walked out. I didn't say a word; just asked her how much I owed her.

She asked if I was paying for all of it and I said "Yes ma'am" and Gary added that he would cover some of it. I told him that I had

It. So I paid her and we went back outside to the truck.

Mick was standing there and asked if we were ready. "Let's Get It" Gary said and we all piled in the vehicles and headed down the road. We passed the garage and Gary's house. Ron was standing out front of the garage talking to someone in a van.

When we went by Gary laid on the horn, stuck his arm out the window, raised his hand a little and with one finger pointed down the road. Ron just waived back and continued talking to the person in the van.

About that time I wondered how long it would be before Mick came flying by us in the Nova. He never did. He just stayed behind us.

We were only about two miles from the store when Gary made a right turn; it was blacktop for about a block then it turned to broken up asphalt, dirt and rock and then just dirt and rock.

We road down the road listening to the radio, by now everyone was feeling pretty good and as usual it was time for the stories to start coming out. Going down the road you could feel the temperature start to drop a little.

We were going down these hills towards "The Creek". That's all they ever called it. We came to a down grade and at the bottom was a bridge, an old bridge built in the forties or fifties.

It was just big enough for two cars to pass; don't try a dually truck and a car, it's not happening without some body damage.

Anyway made of cement and rebar the old fashion way. It was thick for a small bridge. You could tell it was made to last. We pull up on it with Mick. Gary stops and Mick stops behind us. We shut the truck off.

Getting out I noticed Mick was walking back to his trunk. He opened it and you could hear the music, but it wasn't real loud. Gary reached into the bed of the truck, pulled out a fishing pole and handed it to me. Mick then walked up with a bottle of wine. Handing it to me he said, "You got to try this".

"What is it"? I asked

"Blackberry Wine" he said with a smile and took the pole that Gary was handing to him with his other hand.

Taking the bottle Mick had handed me, I took just a little taste. It tasted awesome, definitely not your store bought wine. It was home brew.

I could have drank the whole bottle in ten minutes, just that good is what I told Mick, he replied with "I thought you might like it. Drink Up"!

This time I took a good pull of it then handed the bottle back to him.

In the meantime the rest of the boy's were

setting their poles up to go fishing. Mick and I walked to the bridge side of the truck with our pole in one hand and a beer in the other.

We stepped to the edge and looked over into the creek. It wasn't big but big enough for catfish and that's all that mattered.

The creek was beautiful. Standing there you could feel a slight breeze, it wasn't real hot there. You could hear the sound of the water and the birds. The radio wasn't too loud yet, it was country. Roger and Gary walked up their poles baited, Gary handed me the bait then Roger and him threw their poles in. Baiting my pole was the next thing I did.

"I just love to catfish"

I handed the bait to Mick; and so there we were, fishing, having a beer and talking a lot of B.S about when we were younger and that's when we all heard at the same time. There was a car coming our way from the other side. The motor was screaming and you could hear the rocks hitting the fender wells.

Mick said that it was probably Jimmy and that that is what he sounded like.

I just waited, looking for a glance of that car that was making all that noise and not a bad thing at all. It sounded cool coming down the hill.

Then there it was; My mouth just came wide open, smiling right away and feeling like a little kid. It was a Seventies Gran Torino Sport.

Oh Yeah; The one with two headlights on each side of the grille. The grill was the square type but not real big, and once again; Of course it was a Fast Back.

Later I learned that it had a motor swop from the three-fifty-one C.I.D Four barrel head engine to the built three-ninety C.6 A/T power plant. Color of course: was your

STANDARD YELLOW WITH BLUE STRIPES, A BENCH SEAT AND CENTERLINE WHEELS, BLACK WALLS OUT. PULLING UP TO WHERE WE WERE PARKED OF COURSE IT WAS JIMMY.

HE SHUTS THE MOTOR OFF, STEPS OUT AND SHOUTS, "IT MUST BE PARTY TIME WE'RE ALL HERE"! HE THEN WALKS AROUND SHAKING EVERYONE'S HANDS AND PATTING THEM ON THEIR BACKS TALKING BULL TO EACH OTHER.

THEN HE GETS TO ME, LOOKS AT THE BOY'S AND SAYS, "SAY, WHO'S THE NEW FISH". THEY START LAUGHING AND INTRODUCE ME TO HIM AND WE ALL START TALKING. FUNNY, THEY DID MOST OF THE TALKING; I WAS THE NEW COMER SO THERE WAS A LOT OF LISTENING ON MY PART.

THERE IS A COUPLE REASONS WHY I DO THIS; THE MAIN REASON IS TO FIND OUT WHO THEY ARE FIRST BECAUSE IF YOU LISTEN TO SOMEONE LONG ENOUGH A PERSON WILL TELL YOU WHO THEY ARE AND WHAT THEIR ABOUT.

THE OTHER REASON IS THAT THEY SAY FUNNY

STUFF, ESPECIALLY AFTER THEY HAVE HAD A FEW. THEY TRULY "WANT" YOU TO KNOW WHO THEY ARE.

We had been there for about a hour and a half when Roger said, "Their not biting" and they all kinda chuckled about going back to the house and killing a couple of chickens to go with the greens and beans. We all agreed. We put our poles up and put them in the truck bed. Mick and Roger were in the Nova, Gary and myself in the truck and Jimmy jumps into the Torino, fires it up, drops it in gear and floors it. The tires start spinning and don't stop until he lets off it right after he leaves the bridge.

There's a place to turn around and so that's where we head. When we get there Gary lets Mick go in front of us. Then he tells me; "Watch this";

Mick and Roger pull down to the bridge, stop just on the bridge and then Mick floors it.

The Nova spun its tires all the way across. Gary tells me, "Just something we've been doing forever".

He pulls on the bridge, stops and spins his tires about two feet. "It doesn't really matter how far you spin your tires, just so you do, tradition, ya know"!

"Yes I do" was my reply. So we headed back to the house. The ride was not a big deal this time. Just, Gary and myself riding along, listening to the radio. We were not that far away anyways.

Arriving at the house, of course Mick was already there, trunk open and radio on, Jimmy was there also. He was leaned against his Torino with a beer in his hand and talking to Mick.

Roger must have been in the house, I didn't see him till later and I do not know what direction he came from. Gary pulled up where he had parked earlier that day. We got out and walked over to where Mick and Jimmy were. As I was lighting a cigarette, Gary just comes out with "Let's get to killin them Chickens", I started laughing.

They just looked at me, I couldn't help it. The way he said it and the timing. Eight beers might have had a little to do with it, but it was like I was sixteen again. I'll explain later;

Anyway, so we all agreed and started walking to the barn. Mick says "Hey Gary, Roger said he had two young roosters penned up. Those are the ones were supposed to dress out".

All the way to the barn I was thinking I don't want to have to clean these chickens, I didn't have my chicken killin rags on and plus I had a buzz going on.

We walked up to the front of the barn but the doors were paddle locked so we went through the side gate. It was a nice barn and it still had the tin roof on it. You couldn't see directly through it either.

Around back he had the two penned up; Jimmy just took over at this point. He told Mick to drag the hose around I'll take care of them. That he did. The old ring their neck routine.

By the time Mick came back Jimmy had one half cleaned or skinned should I say. There wasn't any feather pluckin goin on. Roger showed up with a couple of guns and a rifle. Gary had lit the grill and turned the radio on.

The barn had electric and it also had a little area, not to big, but it had a table, some chairs, a couple old tool boxes that the radio sit on and a garbage can made from a fifty five gallon drum and it looked like they might burn wood in it during the cooler evenings. Just enough room that four or five people could play a friendly game of cards.

Mick and Jimmy were cleaning the chickens, the barbeque grill was lit so Roger, Gary and I walked out back of the barn a little. Roger had made a homemade shooting range right there.

Target one, a dead tree, red circle on it, big; Small circle in the middle, yellow. Target two, refrigerator, Target three, water heater and also small targets placed randomly. They had different spacing and footage I don't know.

Anyway, Roger asks me if I shoot much; That's a nice way of asking somebody. Do you know how I told him; I used to shoot and hunt; That I enjoyed hunting rabbits and quail out west.

He handed Gary a pistol, I think it was a 357, me a 22 rifle semi-auto and he had a 38 special I think it was. You could tell they were a little skittish. They wanted to know for themselves that I could shoot before they gave me a big weapon. I don't blame them at all.

Don't get me wrong, I know a 22 caliber riffle or gun can kill you but if you have been around gun's you know what I am talking about. If you don't get some gun training it may save your life someday. Plus it's fun shooting targets.

Anyways we shot off some rounds, challenging each other to shoot some of the smaller targets when Jimmy and Mick showed up.

Jimmy said "Them birds need some attention".

Roger hands him a pistol and heads to the house telling us he needs to get some things that he'll be back. We shoot a little more than move over around the grill.

There were some tree stumps and old chairs around it. you could sit and B.S with the chief. Roger shows up with stuff to cook the chickens and does his thing. We all just sit or

STAND THERE TALKING WHILE THE BIRDS ARE COOKING.

TO ME THERE LOOKS LIKE ABOUT A HOUR AND A HALF, MAYBE TWO BEFORE IT STARTS TO GET DARK. JIMMY TELLS ME IT WOULD PROBABLY RAIN A LITTLE AFTER DARK. IT COOLS THINGS OFF; IT'S WHAT IT DOES AROUND HERE.

EVERYBODY WAS TALKING ABOUT SOMETHING OR SOMEONE THEY KNEW.

WHILE WE WERE TALKING I WAS LOOKING AT THE COUNTRY SIDE BEHIND THE BARN. THEY HAD A LOT OF LAND CLEARED FOR GRAZING. THERE WERE TREES OUT THERE A WAYS. THEY LOOKED THICK ON THE HILL.

THEY HAD SOME OLD CORRALS, SOME HOG PENS AND CHICKEN COOPS; O YEAH THEY HAD A SCRAP IRON PILE. EVERY BARN HAS ONE.

SOME TIME PASSED. THE BIRDS WERE DONE. ROGER SAID COME ON, LET'S GO UP TO THE HOUSE AND EAT. WE ALL STARTED WALKING THAT WAY. WE WERE WALKING TO THE SIDE I COULDN'T SEE.

As we rounded the corner I could see a little screened in porch, just big enough for a big picnic table, benches, and a few chairs; that was all. There was a long narrow table up against the screen.

Getting closer I could see there was food on it. The kitchen was right inside that door. The kids were seated there, or should I say bouncing there.

Mindee walked out, told the kids to settle down in a raised tone. Reaching the porch door Jimmy opens it. Roger carries the birds to the table and sets them down. Roger tells Mindee, "Mines done, yours".

She just smiles and tells him, "What do you think".

Stepping into the kitchen she tells the other girl, I told you earlier I forgot her name, "Come on its ready".

The girl steps back out with a picture of cool-aid and a hand full of cups, plastic that is, poured the Cool-Aid and then sat down. Then we sat down. Mindee asks who is saying grace; Gary spoke right up "I will".

She tells him Thank You.

We all bow our heads then Gary began. At the end of his prayer I felt surprised. I don't know why. I really didn't know him that well. Or maybe just because I hadn't seen that in him just yet, but it was a very good prayer. Seemed like one from the heart just for the occasion. After our A-Men we dug in.

It wasn't a quiet dinner. The kids were talking, if that's what you want to call it, we were eating and talking about this or that. The food was awesome. I made it a point to tell the girls how good it was. Roger had done a great job on the chickens. They were very tasty.

During dinner the sun had went down and it was dark, the house had no booger lights that I could see. During dinner Roger had mentioned that he had a surprise for Gary, Jimmy and myself.

The girls had gotten up from the table and were tending to the kids. Roger and Gary got up, took the platter that had the pieces of chicken on it into the kitchen.

There was very little left, I do mean very little. I asked could I do something to help, they said no. Gary picked up the big bowls of greens and beans, stepped into the kitchen behind Roger.

Jimmy and I started talking about what happens around there, the day to day things. When Roger and Gary stepped back out Roger had a rifle.

It looked old but in good condition and Gary had a box of ammo. Roger says "Let's go have some fun". He had left the other riffles and the pistol down at the barn.

My first thought was does he have spot

lights on the barn that show the targets? As we left the porch it kept getting darker. As we reached the corner of the house I looked down at the barn, All I could see was a light in front that one of them must have turned on when we left.

We reached the barn and went around back. There was a small light back there; also one in the barn over the card table. We all walked inside where the card table was.

Gary opens the box of shells he had been carrying, Roger asks me have you ever seen tracers before; No I replied, he said "Well your fixin to" and they loaded the rifle. It turned out to be an old M-1. Stepping outside with the riffle Roger tells Gary to shut the lights off and he does.

Then Roger starts shooting the riffle. It was too cool. You could see the bullets flying down the field. He also shot the water heater with it. That was cool too. I had seen it done before but every time is still

A RUSH, JUST THE NOISE OF BULLETS FLYING.

THEY WERE RED TRACERS. DON'T LOOK IT UP ON THE COMPUTER, NOT EVEN CLOSE, YOU HAVE TO WITNESS IT LIVE. PUT IT IN YOUR BUCKET FOR SOMETHING TO DO; FOR YOU BUCKET LIST FOLKS. AFTER HE SHOT IT A FEW TIMES HE HANDED THE RIFLE TO ME.

THAT'S WHEN I KNEW THEY HAD ACCEPTED ME. JUST A LITTLE THOUGH, JUST ENOUGH TO SAY YOU CAN COME BACK AROUND AS LONG AS YOU ACT RIGHT AND NOT THE KIND YOU MESSED UP, WE WILL LET YOU SLIDE THIS TIME, THAT TAKES YEARS.

I SHOT A FEW TIMES THEN HANDED IT TO JIMMY. HE WAS THE CLOSEST TO ME. HE DIDN'T WANT TO SHOOT SO HE PASSED IT TO GARY. HE SHOT IT A FEW TIMES THEN WE PUT IT ON THE CARD TABLE WITH THE REST OF THEM.

WE STOOD AROUND TALKING FOR ABOUT A HOUR, JIMMY SAID HE WAS LEAVING. THAT KIND OF BROKE THE PARTY UP. ROGER PICKED UP THE WEAPONS FROM THE TABLE AND SAID,

"I'm going to the house, shut the lights off when you leave, see ya'll later" then looked at me;

"Nice to meet you maybe I'll see you around somewhere".

Jimmy said about the same thing, I didn't really except anything more than that, after all they had opened the door to a day in their lives to me;

If you understand that type of thinking; I replied you never know it's a small planet and the pleasure was mine to meet ya'll.

They faded into the dark as they walked away from the barn. Gary and I looked around at some of the junk Roger had collected. He had some interesting pieces, if you like old stuff that people just throw out.

Most of it just needed a little time spent on it and it would be working again. Gary asks me, you ready to go, I replied "Let's get it".

He shut the lights off and we started toward his truck.

Reaching the truck we got in, Gary cranks it up then asks me, what I thought about the day so far. I responded by telling him this;

"It's been one I will never forget, Thank You". He smiled, looked right in my eyes of the lighted cab and said,

"We were truly blessed today, Know what I mean".

"Yes" I said and he turned the radio on, backed out and headed for his house.

Not a whole lot was said on the way. I think we were just taken a break for a minute, just listening to the radio.

Turning into his driveway you could see there was someone on his porch. Gary said someone's looking for me and he pulls the truck up beside the porch and shuts it off. I could see it was a woman.

Gary asks "How are you tonight?"

She replies with a "come to see you".

Gary told me her name but I have forgotten it. We got out and stepped up on the porch. There was a recliner, brown leather, some chairs, the plastic kind and a coffee table or something like it.

Gary and the women sat down on the plastic chairs so I just went and sat in the recliner. They were talking, I was just setting and looking out at the sky from the porch not really paying attention to what they were saying.

It was nice out and I was also getting tired. Then Gary brought me into the conversation. He was talking to her about the day. We probably talked for about a hour, it seemed like three though.

Then Gary told me where the extra bedroom was, also the bathroom, "we're going inside, I'll see you in the morning".

"I'M SURE YOU CAN FIGURE IT OUT" WERE HIS LAST WORDS FOR THE NIGHT.

I SAT THERE FOR A LITTLE LONGER THEN I RAISED THE FOOT PART OF IT UP. SO NOW I AM LAYING BACK IN THIS LEATHER RECLINER.

GARY HAD TURNED OUT SOME OF THE LIGHTS ON HIS WAY IN. THERE WAS ONE SMALL LIGHT BULB ON. THE BREEZE WAS BLOWING. I COULD HEAR THE LEAVES RUSTLING TOGETHER. IT HAD ALSO COOLED OFF EIGHT OR NINE DEGREES. WHAT DO YOU THINK HAPPENED? EXACTLY; I FELL ASLEEP.

THE NEXT THING I REMEMBER WAS GARY ASKING ME IN A LOUD VOICE FROM INSIDE; "DO YOU WANT A CUP OF COFFEE"? I SAID ABSOLUTELY! I WALKED INSIDE, HE HANDED ME A CUP OF COFFEE, THEN ASKED,

"THAT RECLINER GOT YOU DIDN'T IT"? I SMILED AND SAID YEAH. THEN GARY ASKED ME WHEN I WAS LEAVING, I REPLIED TO HIM IN A HOUR OR SO. HE ASKED ME WITH A SMILE ON HIS FACE,

"WE HAD A GOOD DAY YESTERDAY DIDN'T WE"?

"Yes we did" I replied

"Let me get you a towel and some soap while you go and get some fresh cloths were his next words to me. I walked out to the car, it wasn't wet.

You could smell the trees it was a clear sky day. It was quiet too. I took some clothes out of the car and headed back to the house and opened the door.

He was cooking something. I didn't know what it was but I figured anything would be great especially since I didn't have to cook it. Know what I mean?

I met Gary right before the kitchen and he handed me the things. Off to the shower I was. When I came out Gary was sitting at the table.

He had a pot of coffee, some eggs, spam and some toast sitting in front of him. Something for the road he said. Sitting down at the table I said Thanks.

We were eating and talking about things for about a hour. It seemed like when we talked we always learned more about each other. We were alike in many ways. The day together was interesting. We both knew it was just one of those things that was meant to be and we appreciated that.

Getting up from the table we walked toward my car. He said let's stay in touch and I said Okay. We exchanged phone numbers and hugged.

Sitting down in my car I said "See you later". He replied with you bet and I cranked the car, backed out and right before I turned the steering wheel to the right (which would of pointed me to the road). I then waived and headed out the driveway to the blacktop.

When I reached it I turned the wheel to the left, stepped on the gas and up the road I went. Passing Ron's garage to the stop sign I made a left again and passed the store.

Headed towards Roger's house I smiled to myself.

Inside I was thinking I know people around here. My car being broke down had turned into a good thing. After passing Roger's driveway I realized I was on the road that I had never traveled down. This really started me thinking about the way things are now.

Before I start in on that I should probably tell you a little about me and the way I was raised. Some of it may seem unreal but it's most likely all true. I will have to leave some things out or change it up a little because I do not know who's reading this.

Anyway Believe it or not; Here it is.

PART FIVE:

IT BEGINS

I am four years older than Charlie, like I said in the chapter on Charlie, but I was raised differently. As a small child I traveled all over the United States.

My father was a building supervisor for a major company. They built large industrial plants back in the sixties. My mother, brother, younger sister and I would travel with him.

He would finish one job than the company would tell him where to go for the next one. Packers would come in, pack up the house and move it. Our personal things that is, not the whole house. Most of the houses we moved into were already furnished.

My father liked to drive so most of the time

we would drive to the next place. I saw a lot of our country sitting in the back seat. Well I wouldn't really call it sitting. It was before seat belt laws. (Seat belt laws I'll save my feelings on that for later.)

Anyways I can remember going to seven different schools in one year. It was very strange for me as a child. Nobody I met as a child families did this type of thing and as an adult I know nobody who has been through anything like that. Obviously I didn't have any close friends.

The people we would meet was a short term thing. Most of them worked for my father. A lot of them had no kids. So most of the time, I was around adults.

I would listen to what they said. Most of their conversations were about building things that went into the plants my father was building.

We did go fishing, camping and to little

Caravel's. They were awesome back then. Each one had their own special theme. I do not think they have them anymore but if they do it is my fault for not going.

So go with an open mind and watch the carnies do their thing. You will know why people used to say "Oh they ran off with the carnival".

Only one problem with me being around adults all the time, I learned to talk at a different level or I didn't get any attention.

So when I was in school, it was hard to talk to kids my own age. They were talking about what was going on at a level I just couldn't get.

By the time I met them it was time to move. I had little in common with the kids I met for obvious reasons, plus at a young age I was already trying to find out how things worked and it didn't matter what it was.

So there were times I just didn't care to be

around kids my age. They seemed silly and childish. Imagine that!

My mother was always home. She was always teaching us something new. She had educational books around us all the time.

Part of the reason was because we were in and out of school so much and the other reason was to keep us quiet while we were riding all over the countryside in the back of his new Cadillac.

So academically I was in the top of my class; Socially I was not into kids my age. I would go to school, set in class and absorb what the teacher was saying.

When it came time for recess a lot of the time I would stay around the teacher and talk to them. Playing with the other kids seemed like a waste of time. Why make friends, we were leaving soon anyways.

After school I would go home and wait for my Father to come home. We would talk

ABOUT HIS DAY, MOST OF THE TIME IT WAS HIM TALKING, OR HE WOULD SHOW ME THE BLUE PRINTS.

HE WOULD EXPLAIN HOW MANY MEN WERE WORKING THERE, SETTING UP DIFFERENT MACHINES, SUPPLIES THEY NEEDED, THE TRUCK LOADS OF STUFF THAT CAME IN, THE COST OF THINGS, HOW MUCH LABOR WAS AND THE OVER-ALL COST TO DO THE JOB.

AFTER A WHILE YOU START TO LEARN SOMETHING OR AT LEAST I DID.

WHEN HE WAS TEACHING ME HE WOULD KEEP IT SIMPLE UNTIL I UNDERSTOOD WHAT HE WAS TALKING ABOUT THEN HE WOULD INCREASE THE AMOUNT OF THOUGHT IT TOOK TO COME TO THE ANSWER. THAT'S JUST HOW OUR CONVERSATIONS WENT.

HE WOULD ALWAYS CHALLENGE ME TO THINK ABOUT WHAT HAPPENED, WHAT SOMEONE SAID OR MAKE ME LOOK FOR THE ANSWER TO A PROBLEM. HE WAS JUST THAT WAY.

Even when he took me fishing he would start teaching me right then about boats, why one has a thirty five horse power engine and the other a fifteen, why the different shapes and why the weather had an effect on fish. It was always a learning experience with him. I loved it and still do.

I can remember going fishing this one time with my Father and Grandfather. We had drove to Florida on vacation. We stayed with my Grandfather.

He lived on a river in a house. He called it an old Florida river house. It was not real big. Three bedrooms, one and a half bath, as I remember it was made of part block and part wood.

It was painted white with green colored slim on it in places. To me it looked like a big box.

Inside the house it was very clean. The furniture looked real old but it was in excellent condition. As a matter of fact everything looked old. Later I figured out they had Tens of Thousands of Dollars in antiques.

Out back of the house they had a huge porch, it too was awesome. Grandpa called it a Florida room. There were rocking chairs made with real wood; Real nice ones.

There was a slider, steel of course, small end tables made from cypress trees, artificial vines and fishing nets hanging from the roof, lamps that were little, boats, ropes and rugs.

He had anchors, shells and other things hanging on the walls and in the back there was an old cast iron wood heater-stove that had crazy shaped legs on it.

When you stepped out the door there was a path, part of it you could walk on flat blocks, the rest was grass and small rock. It led to a reddish colored looking dock made

FROM OLD TELEPHONE POLES.

It had to be strong because it was holding a big yellow boat out of the river and it had benches on it with a small table.

There was also a fishing station, that's what he called it, looked like a sink to me. There was a big cypress tree on the right side and the left side of the dock looking down the river was covered with trees, vines and dead branches.

There were parts that looked like narrow mud or grass slides. Grandpa said that's where the gators live. Then I walked over to the front of the dock and looked over into the river.

Before I tell you about the river I forgot to tell you why my Grandpa called it a Florida room. The windows were pieces of glass, all the same size and assembled in metal frames.

You could roll them open or shut. Open was better or least for the time that I was there.

Smelling the trees, flowers and the river was great. The breeze in the evening was good plus no mosquitoes. They could keep the rain out also. He called them jealousy windows.

Anyways, it was clear with a very little brown haze or stain. Grandpa told me the haze was from the rotting tree leaves, branches and stumps. You could see all that along with the green vegetation.

The bottom looked sandy with brown and gray clumps. The best thing though was the fish. You could see them swimming. It was cool. I still look in every creek, river, lake or where ever there is water to see if it's alive.

That was the first time I had ever saw a river so alive. That was also the moment I fell in love with water that flows, streams, creeks, rivers, You get the idea. The first fish I saw swim by, my Grandpa called a Bass. It was a medium green and white fish.

There were little fish he called stump knockers, they were colorful. There were catfish and turtles too but the one that still sticks out in my mind is the Alligator Gar.

It was five or six feet long and it looked like it was prehistoric. It had a long bill about a foot and a half long with lots of teeth. It had a narrow body with a big tail. It was chasing the little fish.

At that moment all I could think was when can I start fishing. So take someone fishing, start with a kid, you never know the effect it will have on them. It may be just what they need to make it through a hard time.

(I taught my girls when they were real young and they still love to fish. Get them away from computers and cell phones. It's okay to get a little dirty and sweaty. It builds character.)

We had arrived late and it was getting

Dark and the mosquitoes were starting to swarm; sucking the blood right out of us. So we went back up to the house for dinner; catfish, salad and french-fries. It was tasty.

Grandma played the organ; we were all amazed at how good she could play. She could play all kinds of music on it. Then my dad and Grandpa said if I was going fishing with them in the morning I had better go to bed and get some rest.

Now looking back that must have been real funny because there was no way I was going to be able to go right to sleep after them telling me we were going fishing. All I could think was I wanted to catch a big one and how was I going to accomplish that.

Oh yeah, the big yellow boat was on my mind as well. It looked like pure fun to me. They woke me up the next morning and my first thought was riding that boat. We had a small breakfast, I couldn't stand it, I was so ready to go. So when he said "Let's head out"

All I could do was smile.

We stepped out the door and headed for Grandpa's truck. I thought, what's all this about, but I never said a word and I just followed them.

They had loaded the truck with poles, a cooler, one bait bucket, a net that you throw to catch bait Grandpa said, a hoop net with a long pole, two tackle boxes and a black plastic bag with a little something in it.

On top of this, it was kind of raining; you know the kind that wet's you very slowly. So we jump in the truck, crank it up and started for the road.

We didn't even make it out of the driveway when the education started on fishing. They were both going at it.

The weather, sun light or lack of it, the tide, why you fish on the outgoing tide at the place we were going, live bait or artificial,

WHEN TO USE THEM AND WHY, WHY WE HAD TO STOP AT THIS BAIT STORE FOR SQUID WE NEEDED FOR PIN FISH AND IT JUST WENT ON AND ON.

WE FINALLY GET TO THE BRIDGE. IT WAS A BIG BRIDGE WE WERE GOING TO FISH UNDERNEATH IT SO WE WERE OUT OF THE RAIN. THAT WAS A GOOD THING BECAUSE IT STARTED TO COME DOWN HARDER.

I COULD SEE FOUR OR FIVE PEOPLE FISHING. GRANDPA PULLED UP, HIM AND DAD GOT OUT, I WAS RIGHT BEHIND THEM AND THEY HEADED FOR THE PEOPLE; NEVER EVEN TURNED TO GET A POLE. "WHAT'S UP" I HAD THOUGHT.

WHEN WE GOT OUT THERE, IT WAS A LARGE OPENING RIGHT INTO THE BAY. IT WAS AN OUTGOING TIDE AND IT WAS MOVING FAST. YOU COULD TELL BY LOOKING AT THE PYLONS ON THE BRIDGE.

THEY TALKED TO THE PEOPLE FOR ABOUT TEN MINUTES THEN GRANDPA SAID LETS GO, GET THE POLES, I THOUGHT FINALLY.

As soon as they got out of ear shot of those people my Grandpa said "They have no idea what they're doing" and then chuckled, so did my dad.

Back at the truck they took three pole's out that looked alike out of seven or eight. I thought that was odd, took out just one tackle box, the hooped net on the long pole and that was it.

On the way back to where we would be fishing they were talking about the size and color of artificial bait we would be using.

We stopped at a place and they told me to fish there, they tied a lure on a pole and for about fifteen minutes showed me and explained to me what to do. Then they moved a few yards from me. I saw them talking but I couldn't hear what they were saying.

They had opened the tackle box and were pulling out lures. My dad shook his head no and my Grandpa smiled and pulled out two

LURES AND HANDED ONE TO MY DAD. HE TIED THE OTHER ONE TO HIS POLE.

I WATCHED THOSE TWO MEN CATCH FISH FOR TWO OR THREE HOURS SO DID THOSE OTHER PEOPLE. LONG STORY SHORT; ON THE WAY HOME I LISTENED TO THEM TALK ABOUT WHY THEY WERE CATCHING THE FISH.

LATER THEY ASKED ME IF I HAD LEARNED ANYTHING. I DID, BUT THE MOST IMPORTANT THING I LEARNED THAT DAY HAD NOTHING TO DO WITH FISHING.

IT TOOK YEARS FOR ME TO UNDERSTAND WHAT I HAD LEARNED THAT DAY BUT EVEN AS A YOUNG BOY I KNEW MY FATHER AND GRANDPA HAD A SPECIAL CONNECTION BETWEEN THEM. SOMETHING MORE THAT FATHER AND SON AND I SAW IT.

BACK ON THE HOME FRONT WHILE WE WERE ON VACATION MY FATHER APPARENTLY DECIDED TO CHANGE JOBS. I HEARD WHY AND MAYBE LATER I WILL PUT IT ON PAPER.

WE WERE STILL LIVING OUT WEST AND MY DAD

was working for a big company. Everything seemed to be going well. Our family seemed to be happy and prosperous.

I would hear my parents and their friends talk about the importance of family and friendship. We did things together; we even went to church sometimes.

Life was good. Than my mother went to work at a hospital. This was strange to me because up to this point she had always been with us kids at home.

This went on for awhile when I began to notice things changing between my parents. Arguing became frequent and this went on for awhile, how long I don't really know; I just tried to tune it out.

Then one day they said we were moving to Florida. That didn't bother me at all. I was use to moving but not the way we did this time.

There were no movers. My dad had quit his

Job so we loaded a trailer with all our belongings and one Saturday or Sunday morning, I don't remember which but we went to the Park-N-Swap or flea market whatever you want to call it, sold everything, took the trailer back to our friends house and dropped it off.

While we were there we ate dinner. They also had some puppies. I wanted one. They were golden retrievers, I don't remember how we ended up with one but we did. It was a girl and we named her Peppermint Pattie from the cartoon Charlie Brown.

After we ate we took the puppy, got into the car and started for Florida. It seemed like we drove through the desert for days. There were few trees and it was super hot.

The towns were spaced far apart, the big ones that is. There was gas stations maybe some type of store, probably family owned in between.

I believe it was nineteen sixty nine; I-10 was the road we were on and I do remember riding for hours and hours through towns that had been destroyed. There were ships setting in the middle of these huge houses and buildings.

Cars, trucks, furniture, stoves beds and there was garbage everywhere. I saw upside down buses, the trees were broken down and lying all over and people were cutting things up that were in the roads. I remember being scared.

My father was talking about gas for the car; he said it might be a problem finding some. My mother really didn't say anything she just kept looking. Now I realize she probably was in shock.

We kept going until we finally came to a bridge. There was a police officer. My father got out and went to talk to him. Apparently, he had told my father to cross at his own risk. He did slowly.

There was devastation everywhere. It was freaking my parents out. I could hear them talking about how many people might be dead, the wreckage, how bad it was and the effect it would have on the people who lived there.

They said a prayer for the people and they talked for hours. I just held the puppy, wondering what had happened. Then after awhile they just stopped talking about it.

After a bit I asked what had happened. They said a hurricane had passed through. The hurricanes name was Camille. I asked him what a hurricane was and he told me everything he knew about them.

The next thing I remember was arriving at Grandpas house. Everything else seemed boring after passing through an area where a hurricane had been.

I have since in my life seen where tornadoes have passed but the one thing I never have

Forgotten about with that hurricane is the ships were not in the water, they were in the middle of the houses, or on top of cars, semi's, or whatever they landed on. I am talking about ships, one hundred feet or bigger.

So we stayed there until we rented a house. It wasn't that far from my Grandpas house. It was cool to me having ponds close by, a lake that we went swimming in, you just walked up to the entrance, paid him fifty cents and went in. Inside was nice. It had a large white sand beach.

There was a dock off to the left. There was a ski boat there and they were teaching people to ski. I used to watch them start from the dock. That was interesting to me.

To the right there was a concession stand, they sold hot dogs, hamburgers, fries and soda from, the usual but the best thing was they sold deviled crab rolls. I still can taste them.

It also had a covered area, screened in too. It worked great for getting out of the sun. There was always music playing in there.

Looking from there to the lake there was a platform off the beach a ways. It had two diving boards on it. One was just off the water a little the other was about twelve feet or so off the water. There was a slide as well.

In front of the platform fifty feet or so was a big floating platform with nothing on it, you just swam out there and sunned yourself.

You know like a seal. I didn't see much fun in that but I would swim out there and get on it for a minute or two then jump off the diving boards.

There were also volley ball nets. What's a beach without some volley ball really? They had other activities but my point is; you got all that and also a life guard on duty for

Fifty cents.

So I spent as much time there as possible. I met some kids there and we would fish those ponds. We did catch fish, a lot of big ones and small ones, all types.

We were not there long though, my parents bought a house about fifteen miles away. It was small but they had grand plans for it. I also met a lot of kids in this neighborhood.

Some I have known for years now. We lived there for three or four years. Everything seemed good. My dad was working for a company that made fiber board or something similar to it.

At some point he started selling cars with my Grandpa. He seemed to be happy and so did my mother. We threw pool parties; (You talk about stories; I have a few about that pool) We rode dirt bikes, fished and camped. We were having a good life, or so that's

WHAT I THOUGHT.

I REMEMBER MY FATHER ALWAYS TALKING ABOUT OWNING A BUSINESS TO ANYONE WHO WOULD LISTEN. APPARENTLY MY MOTHER DID NOT LIKE THIS IDEA BUT AT SOME POINT HE WENT FOR IT.

APPARENTLY IT CAUSED PROBLEMS AND PLUS, HE WAS DRINKING MORE. THE HOURS HE WAS GONE WAS GETTING INCREASINGLY LONGER. MY MOTHER GOT A JOB AND APPARENTLY THEY COULD NOT FIGURE IT OUT.

I COULD GO ON ABOUT WHAT I HEARD AND SAW BUT THERE'S NO POINT. THESE TWO PEOPLE JUST COULDN'T PULL IT TOGETHER FOR THE SAKE OF THE FAMILY. THEY DIVORCED. SOMETHING THEY THOUGHT WOULD FIX THINGS OR CHANGE THEIR CIRCUMSTANCES. SOMETHING I KNEW NOTHING ABOUT.

I REMEMBER THE SONG, THE LADY ON THE RADIO SANG, "OUR DIVORCE BECOMES FINAL TODAY". HEARING IT DEVASTATED ME. IT ACTUALLY MADE ME PHYSICALLY SICK AND I BROKE OUT IN RASHES.

How could two people who loved me, who taught me things I know and who hauled me all over the United States do this? It was like a bomb.

One minute we are all together, the next scattered all over. What it did to my younger sister, brother and I was separate us and that's all I will say.

I was a very angry person into my adult years. Bottom-Line, I went to live with my father. There was no way my mother could have dealt with me.

I know what it did to my father; I had the front row seat. Watching him fall apart was something "unbelievable". The things I saw, Some of them I was involved in, should have never of happened, but they did.

You talk about eye opening; You have no Idea yet. I know that's why we went to Alabama. He needed to be with his friends because they were also his "Life", his family.

He didn't want the people he knew in Florida to be a part of it.

All my schooling could of never prepared me for the education I was about to get and the reason is because he interacted totally different then I had ever seen. He leaned on these people and they accepted him totally drunk, out of control and broke.

Although he worked, it was just enough to pay the rent and buy a little food, a lot of beer and wine. He worked as a body and paint man. He was very good at it. The man he worked for paid him well when he was working.

Waite a minute; I forgot to tell you about the trip to Alabama. We took a friend of my dad's. His name was Don. He was also going through a divorce and just this side of insane.

He would drink a bottle of vodka a day plus wash that down with beer, smoke three packs of cigarettes and hit a few joints, if you know what I mean.

My father would drink and smoke with him. They would go from the time they got up until the time they passed out. They slowed down when they were working. They both worked at the body shop.

We lived in a three bedroom one bath mobile home in some trailer park. I was thirteen or fourteen watching these men fall apart. Just the ride up to Alabama was eye opening.

We loaded up an old Chrysler Imperial, Nineteen Sixty Eight, I believe. Everything we owned fit in the trunk but it was a huge trunk. You could put half a Datsun 240 Z in there. We filled the gas tank up and headed out.

Oh Yeah, we also picked up a case of beer, two bottles of Vodka, a pound of Bologna, a loaf of bread, Chips, mustard, and a few

SODA'S FOR ME.

THERE WERE TWO GALLONS OF WATER IN THE TRUNK BUT THAT WAS IN CASE THE CAR OVERHEATED. IT WAS CLEAN ENOUGH TO DRINK THOUGH IF NECESSARY.

A LITTLE LATER DOWN THE ROAD I FOUND OUT DON HAD A FOUR FINGER BAG OF POT WHICH THEY STARTED SMOKING ONCE WE HIT THE INTERSTATE.

WE LEFT IN THE MORNING TIME WHICH MEANT THEY WERE NURSING THEIR HANGOVERS. A CUP OF COFFEE AND A FEW JOINTS GETS YOU GOING NO PROBLEM. I DON'T REMEMBER WHAT I WAS DRINKING OR EATING, SHOCK HAD SET IN.

I HAD NEVER SEEN MY FATHER ACT THIS WAY AND FORTY MINUTES LATER THEY WERE DRINKING VODKA AND WASHING IT DOWN WITH BEER AND TALKING ABOUT THEIR WOMEN PROBLEMS AND OTHER B.S. I DIDN'T WANT TO LISTEN TO.

THE ONE THING ABOUT THE OLD IMPERIAL WAS THAT IT HAD A GOOD RADIO. THE SPEAKERS WERE GOOD TOO. SO I BEGAN TO LISTEN TO EVERY SONG.

I memorized the lyrics and who sang them.

It was my way to drown out what they were talking about. I am also pretty sure I was buzzed from all the smoke; If you know what I mean. This was all happening while driving sixty or seventy miles per hour (mph) up the highway.

We stopped a few times to use the restrooms, it was day light outside. We stopped in Cordell Georgia for gas. This is where it started to get a little scary, they filled the gas tank up, checked the oil and water, bought another case of beer then headed west towards Richmond.

It was a two lane that winded up and down when you go over the bridge you will go passed miles of pecan trees. After that it's farm s and small towns.

By now they were feeling no pain. The radio was a lot louder and the car is still at interstate speed. It would have been nerve racking for an adult that was sober. Wonder what affect it would have on a kid!

He said we were only a couple hours away; that it would be the next time we stopped. It had turned dark, we were in the woods, headlights on bright, they were drunk and high, radio was loud and we were going way to fast if you ask me.

You talk about intense; I didn't have time to think about what was happening to me. I was looking for the bad crash we were surly going to be in. Somehow though by Gods grace if you must know; we made it to my Aunt and Uncles house.

They were good people. They saw right away what was going on and they watched over me as much as possible. They also talked to me about what was happening and after a while I learned to love them. "They were real".

We stayed there for about a week then

moved into the trailer I was writing about. Like I said they worked enough to pay the bills which made them late sometimes and then the landlord would show up and have to talk to them. This really messed with my head but after words they would pay on time for a few weeks. Then it was right back to the same thing.

They played poker and this is where I learned how to make some money. In the right game it's easy. Just depends on what's happening during the game.

I watched my dad school these people; He just needed to be a little sober. He knew how to deal the cards plus control the game with conversation. They weren't paying attention and he played to win; it didn't matter how and he was good at it.

Sometimes he had a partner. He was an African American man named George (can't say "Black" anymore) and they would play losers-winners. In which they would split

THE MONEY DOWN THE ROAD AFTER A GAME.

AFTER SEEING THEM DO IT SO DID I IN A PLACE I SHOULDN'T HAVE BEEN. THERE WERE A LOT OF THINGS GOING ON AROUND ME, NOT JUST GAMBLING, DRINKING OR WOMEN SELLING THEMSELVES, A LOT OF CRAZY STUFF AND SOME VIOLENT.

THE MAN WHO RAN THE HOUSE, HE WAS CALLED SHACKLEFOOT. HE HAD THE MARKS TO PROVE IT. I HEARD A FEW STORIES ABOUT HOW HE RECEIVED THE MARKS BUT I NEVER ASKED HIM PERSONALLY. I COULD NOT EVER GET UP THE NERVE.

HE MADE SURE NOTHING HAPPENED TO MY FATHER OR ME. HE KNEW THE PEOPLE MY FATHER KNEW. SOME OF THOSE PEOPLE WOULD HAVE TAKEN HIM OUT AND BURNED DOWN HIS HOUSE JUST FOR FUN AFTER A THREE DAY DRUNK.

BOTTOM LINE; WE ALL GOT ALONG; WE RESPECTED EACH OTHER FOR WHAT WE WERE. SOMETHING YOU DON'T LEARN IN SCHOOL, ITS REAL STUFF.

Sometimes I would get out of school, go home and just wait for them to show up. They would pick me up and we would go out until three or four o'clock in the morning. It was sleep for a hour, get up, take a shower, eat whatever there was and then walk to school. I never knew what was going to happen.

I remember one Christmas; him and Don came home and I was listening to the radio.

(This was about the time I started drinking myself. Not your average two or three beers but a six pack or I wouldn't even get started)

They were talking about opening a garage; I was listening to my father tell Don how this was possible even though they had little cash.

Watching Don's expression I could see that he had no idea what my father was talking about. Even at my age I could hear him talking about the hours, low wages to start

AND THEY WOULD HAVE TO SLOW DOWN THEIR DRINKING OR PARTYING.

I THINK DON JUST WASN'T READY FOR ALL THAT. ANYWAY I BECAME BOARD LISTENING TO MY FATHER TRY TO CONVINCE DON. I WENT TO BED. I REMEMBER WAKING UP IN THE MIDDLE OF THE NIGHT TO A BUNCH OF RACKET; NOT HEARING ANY SCREAMING OR CURSING. I JUST DECIDED TO STAY IN BED.

I WOKE UP AND WENT INTO THE KITCHEN TO GET A DRINK EARLY AND LOOKING FROM THERE TO THE FRONT ROOM, WHICH WAS LIKE ONE FOOT FROM THE KITCHEN IN A PREFAB HOUSE HA-HA!

I COULD SEE WHAT LOOKED LIKE A PYRAMID OF BOTTLES AND EMPTY BEER CANS TIED TOGETHER AND HANGING FROM THE CEILING. I JUST TURNED AROUND AND WENT BACK TO BED AND NEVER THOUGHT ANY MORE ABOUT IT.

AFTER A WHILE WE WERE ALL UP AND IN THE FRONT ROOM THESE TWO MEN ANNOUNCED TO ME THAT THE BEER BOTTLES AND CANS HANGING FROM the ceiling would be our Christmas tree, after all it's the thought of Christmas; Knowing what it's really about; that's what counts.

They wanted to know what my thoughts were. I just remember smiling at these two insane men. I don't know what I said, we just went on with the day but that was our Christmas tree that year.

This kind of craziness went on for about two years then one day we just loaded up in an old car and moved back to Florida and when we arrived back there my dad had met a woman who had some money from a life insurance plan.

Her husband had passed and she was heartbroken and completely going insane. She had been married for a long time. She was lost in this world by herself. That's what she would say. That was an understatement from what I witnessed.

Her and my father started a little used car lot with her money of course. They were spending that plus clothes, vacation and alcohol and lots of that. The money from that first check was gone before the second arrived.

She wised up quick and banked the second one. So the party was over. They sold all the cars and parted ways, happy and glad to have met each other.

We saw her later, she was working and doing fine. That's it: that's all we saw of her. My father had met another woman shortly after that. I won't write about all the crazy things we did before he married her but there were quite a few.

We all moved back to Alabama and he opened a paint and body shop. One of my uncles built engine for hotrods there also.

He was good and I learned a lot from him. One of the more important things was how to tell bad parts from good parts. Let me tell you how I learned this.

My Uncle and I were working on this motor, a four o six — three deuces. It was going in a sixty six Thunderbird. It was going to be my driver.

(I'll tell that story some other time) I asked him to let me build a motor, he smiled and said okay.

Pointing to a pile of parts he had put in a metal box; "Go clean them up real good then wash out the three twenty seven block over in the corner then we will start" he told me.

Telling him "Thank You" I ran over and started cleaning; and I do mean cleaning. They were shinning when I was done. It took me a day and a half to finish.

I couldn't wait for him to show up. I wanted him to know I was serious and could do the job.

He showed up and as soon as he reached the garage I started in talking to him about the parts, showing him what I had done and explaining to him how I cleaned them and so on.

He just laughed; I was enraged. How could he laugh, they were spotlessly clean. Then he told me as soon as you learn the good parts from the bad ones we will build an engine.

I had cleaned a bunch of junk parts. That's when I felt stupid and just turned and walked away, embarrassed.

Later he found me, we walked over to the parts and setting there for hours he went over every part, showing me what was wrong with them. He explained why they were that way and you could see perfectly what was wrong with them. After all they were sparkling clean.

Later we built motors together. For years he told that story whenever I was around. It was the last thing we talked about before he died.

He was truly "one of a kind" and he is Loved and Missed by many; Especially by his best friend and brother; My father. He talks about him all the time. They called him Popcorn.

Any ways, things went good for a while. They started bootlegging out of the garage and that's the time when things began to get crazy.

(I have a lot of stories from these years, I will write only about one and then move on, Save the rest for later on, maybe I will write a book just on Alabama, the statue of limitations may have not run out yet Ha-Ha)

So anyway, at the garage we sold cars and trucks or whatever he took in trade. For work he had done or booze he had used to barter for things.

No money; Bring your mothers broke down

TRACTOR OR WASHER, IT MADE NO DIFFERENCE. HE WOULD TRADE IF THE DEAL WAS GOOD FOR HIM. IF IT WASN'T, YOU WERE "SHIP OUT OF LUCK".

ONE TIME HE HAD TO GO TO ATLANTA GEORGIA, I DON'T REMEMBER WHAT FOR BUT I AM SURE IT WAS ABOUT MAKING MONEY, HE LEFT ME NONE WHEN HE WENT. I WAS OFF SOMEWHERE WHEN HE LEFT FOR HIS TRIP.

HE WAS GONE FOR ABOUT THREE DAYS WHEN I NEEDED SOME MONEY. ONE OF THE GUYS THAT WORKED FOR HIM NEEDED SOME MONEY ALSO.

THERE WERE NO JOBS COMING IN TO THE GARAGE THAT WOULD BRING BIG MONEY AND DAD HAD TOLD EVERYBODY HE WOULD BE GONE FOR A WHILE.

SO ME AND LONNIE STARTED DRINKING ON THIS DAY. IT WAS ABOUT ONE OR TWO O'CLOCK, WE WERE TRYING TO FIGURE OUT HOW TO GET SOME QUICK MONEY.

LONNIE CAME UP WITH THE IDEA TO HAVE A POKER GAME. THAT WAS NO SURE FIRE WAY TO GET

THE MONEY WE NEEDED.

AT ABOUT FOUR OR FIVE O'CLOCK THAT DAY; I HAD WALKED OUT TO THE FRONT OF THE GARAGE. OUT FRONT THERE WERE TWO VEHICLES, AN OLD TRUCK AND A CAR.

I WAS STANDING THERE FOR ABOUT FIVE MINUTES WHEN AN IDEA CAME TO ME.

I TURNED AROUND, WALKED OVER TO MY TOOL BOX AND PULLED OUT MY HAMMER. I THEN WALKED AROUND TO WHERE LONNIE WAS AND BEGAN TO TELL HIM WHAT WE WERE GOING TO DO.

HE MENTIONED TO ME THAT MY FATHER WOULD NOT BE HAPPY ABOUT WHAT I WAS GOING TO DO. I REALLY DIDN'T CARE BECAUSE AT THAT MOMENT I NEEDED MONEY TO PAY SOME BILLS AND THINGS.

SO ON MY WAY BACK OUT TO THE FRONT I PICKED UP A FIVE GALLON CAN OF GAS, IT WAS HALF FULL GIVE OR TAKE A QUART. LONNIE WAS RIGHT BEHIND ME.

WALKING UP TO THE TRUCK I TOOK THE HAMMER

TO THE WINDOWS; BREAKING THEM ALL AND POURED A GALLON OF GAS INTO IT. I TURNED AROUND AND DID THE SAME THING TO THE CAR.

I THEN TOOK A PACK OF MATCHES OUT OF MY POCKET, (REMEMBER, YOU USED TO GET THEM WHEN YOU BOUGHT A PACK OF CIGARETTES, THEY WERE FREE) LIT ONE AND THREW IT INTO THE CAR AND THEN THE TRUCK.

THE FIRE WAS AWESOME, I WALKED BACK TO THE GARAGE AND GOT A CHAIR OUT, SAT AND JUST WATCHED THEM BURN. AFTER A WHILE THE SUN STARTED TO GO DOWN.

PEOPLE WERE DRIVING BY REAL SLOW. I WATCHED THE SUN GO DOWN AND THE LOOK ON PEOPLE'S FACES. IT WAS GREAT; QUITE THE MOMENT.

LONNIE DIDN'T KNOW WHAT TO THINK.

AFTER A BIT THE FIRE DIED WAY DOWN AND THAT'S WHEN I DECIDED TO GO TO BED.

WHEN I ROSE FROM BED THAT MORNING I WENT OUTSIDE, LOADED THE TRUCK ON AN OLD FLATBED truck we used to haul scrap cars to the scrap yard and off Lonnie and I went.

We took them both into the scrap yard and they paid us. We drove to the nearest store, bought a six pack and headed to the garage smiling the entire way there because we had money.

Two or three days later my Dad show's back up at the garage. We are all talking about what had happened while he was gone when he asks about the car and truck.

I told him and his reply at first was that they both ran and could have sold for more. Then there was silence.

I think he was in shock.

He said a few things, I don't remember what they were but they made no difference. The vehicles were gone and so was the money. He wasn't real mad but he wasn't real happy either.

He just chalked it up to a life experience I guess and we moved on. Later he would laugh and tell the story many times.

The woman he had married started having a lot of mental problems; so back to Florida they went. I didn't go right then. My father told me they were going back one afternoon.

He was going south and I was going north on this road. He flashed his lights and we pulled over. He got out, so did I. He began telling me I knew it was the truth.

He never really told me a lie about serious stuff. I told him I didn't want to go. He tried explaining to me that I was too young to be there by myself, it didn't matter; I had other things happening.

Life was happening to me in a spectacular way at the time. (I'll explain later, but not in this book)

My father had left and I had to find a job. You can only eat ninety nine cent pizza's for so long. I went to work for a lumber company, working the green end of an edger. I was catching two by fours, two by sixes and stacking them in piles for the fork lift to pick up.

He would then put them on a semi trailer bound for a sale yard. But before I go too much further; Let me tell you how I got this job.

I just walked up to the man who owned the company and asked him. He had already known my father and me so, No problem.

He said yes, then told me to go tell the girl in the office that he had hired me. She would be the one to give me a paper to fill out, so I did.

She gave me a paper and I filled it out to the part where it asked me for my social security number. I didn't know it!

What was I going to do; it had to be on this paper when I turned it in.

My solution was simple. I told her I had forgotten it. She just smiled. I looked young and that's because I was.

I am sure she just figured I had not memorized it yet; I would turn it in the next day. Her reply was a simple "Okay".

I left there and went straight over to a friend's house and I told him my dilemma. He just laughed and told me for a case of beer in which I could help him drink, he would solve my problem.

Back from the beer store we opened a beer then he said; "Just use mine" and really started laughing.

He explained to me social security in a way I have never forgotten. I turned it in and everything was fine. Don't try this type of thing in today's times. It's called identity theft.

Plus, they want a small book on you before they hire you and that's if you don't have to apply on-line; Oh Yeah, don't forget the five forms of I.D. that's required.

Anyways, I worked there for a while until I got into a little trouble. Nothing real bad but the sheriff told me I had better think about going to Florida to be with my dad. It was his way of telling me; the party was over.

So I packed up my car, a two door Buick Skylark; a nineteen-sixty's model with a Nail Head V8 motor then started towards Florida.

There are a lot of stories I could write about the time I spent in Alabama by myself. It was crazy and scary but I could taste and smell life; It was awesome.

So on my way back to Florida I was riding along, listening to the radio, everything was fine until about Brooksville.

That is when I started hearing a noise. I was already good at fixing cars so I knew

Something was wrong.

Pulling the car off the highway onto the shoulder I could smell something. I shut the car off and walked around to the passenger side front tire. I could see some smoke, nothing real bad but it was still a problem.

Not wanting to drive the car any farther, afraid of some serious damage to the spindle, I went back, opened the trunk and pulled out my old bumper jack (if you remember them you got to be over 40).

Walking back to the front I put it in the slot of the bumper then started jacking it up.

Once the tire cleared the ground I knew what the problem was and plus the dust cover was off

Apparently when my friend packed the wheel bearings he didn't put the cotter pin in to hold the nut that tightens the wheel bearing in place. It had backed off.

Trying to tighten the nut I found out right away it was stripped. There was no way to tighten it.

I thought for a minute on it; Well, probably thirty when an idea came to me. I'll just put a washer in front of the nut and use a cut nail.

Driving the cut nail through the hole in the spindle would hold the wheel bearing in place and I could roll down the highway.

Only one problem; The washers I had where to thick and the hole in the washers were just a hair to small.

Re-evaluating; I figured out the proper thickness would be about three washers leaving the nut off.

So with a rat tail file, I made the holes just a little bigger. Having a can of wheel bearing grease in the tool box I packed the bearing, put the washers on, took the cut nail and pounded it into the spindle.

Shaking the wheel and spinning it to check the tolerance; Ha-Ha it looked good.

I let the car down and put the jack back plus the tools, jumped back into the car and eased it down the road, listening to every noise that car made.

Trying to smell smoke or see it and pulling over every five or six miles at first. It was going fine. A blessing!! I drove it all the way to Tampa.

When I arrived at my dads' house I walked in the front door. I could tell by the way he talked and looked that things had gotten worse. How bad I was not quite prepared for.

I was there about a week or so, it was a Friday evening and it wasn't real late when his wife committed suicide. That's all I'm going to say about that.

Another tragic event had occurred in my father's life. He kind of lost it again; Imagine

That! During the time I stayed there my dad had secured a job for me working at a car lot he was working for, if that s what you want to call it.

I was making One Hundred Fifty Dollars a week and he gave me a full tank of gas every week, him and I moved into a two bedroom duplex across town.

He was never there he was always out going nuts. We weren't there long he had just quit coming home. I might of seen him once every two weeks. He stopped paying the rent.

The man who owned the duplex came up to me and told me I had to move. No big deal I was working so I moved into an efficiency place. It cost fifty five dollars a week. Things weren't bad.

During this time of my life I hung out with a group of kids I had spent a lot of time with when I was younger. We did a lot of crazy things. Let me tell you a couple before I move on.

We lived by a college; don't really remember the name but this college had an automated lunch room. Everything was in machines. They were just building the college, it had portables. They were 12 x 34 or something like that.

Anyway, inside were all these machines, sandwiches, sodas, soup, vegetables, popcorn, potato chips and candy bars. It was the first time we had ever seen anything like it.

Oh, there was a microwave; that too was something new to me. It was very large. Three feet long maybe a foot and a half deep which was big enough to microwave my favorite thing at the time.

Getting a Ham and Cheese sandwich out of the machine I would set the dial to sandwich, push the button and in a minute or so the cheese was melted. The sandwich was awesome.

That's all I can describe to you; Modern technologies; Cool.

It's wonderful being a kid. I couldn't see what that would do to our society and its health. The family dinner was changed forever. Pull it out of a packet or a box and just nuke it. (That saying came later)The corporations Loved it though.

Big money was saved by getting rid of cans, plastic or boxes; just add water and done in five minutes. Sit down and wait for the noise, whatever it was to tell you it was done.

Did I mention that there was a dollar bill machine? Well there it stood about five foot tall and three or four feet wide. Maybe two feet thick; Free standing not even bolted to the wall.

After a month or two of eating at the

LUNCH ROOM WE ALL FINALLY NOTICED THIS FACT.

I DON'T REMEMBER WHO MENTIONED IT FIRST BUT WE ALL AGREED ON ONE THING. IT MUST HAVE HUNDREDS OF DOLLARS IN IT AND WE HAD NONE. I HOPE YOU CAN SEE WHERE THIS IS GOING!

IT WAS ABOUT TEN O'CLOCK THAT NIGHT; SUMMER TIME TOO. IT WAS NICE OUT AND WE WERE ALL AT THIS PARK WE HUNG OUT AT.

WHAT ABOUT OUR PARENTS YOU ASK; DO THEY KNOW OR EVEN CARE? "STOP" SOME OTHER TIME ON THAT.

SOMEONE MENTIONED A DOLLAR MACHINE. SO TO THE COLLEGE WE ALL GO. WHEN WE ARRIVED, WE ALL SPREAD OUT IN DIFFERENT DIRECTIONS AND FOUND OUT WHERE THE GUARD WAS. IT TOOK ABOUT TEN MINUTES AND THEN WE ALL MET BACK AT THE PORTABLE.

WE SET UP A PERIMETER SO THAT NOBODY WOULD GET CAUGHT INSIDE THE PORTABLE. TWO WENT INSIDE AND ABOUT TWO OR THREE MINUTES LATER CAME BACK OUT AND WAVED THEIR HANDS to leave and we did.

Meeting back up down the road they said it was too heavy to carry but that wasn't stopping us. Now it's a challenge! Forget about the money!

We started thinking, could we use a wagon? Not strong enough somebody said. What about a piece of plywood and some wheels? Somebody's dad had one like that. No: It might fall off one side or the other. We need to strap it down.

Then the answer came; Someone knew where there was a hand truck. They used it to move refrigerators. It also had straps; Yeah, Game On!

Taking the hand truck we all started back, found the guard and set up perimeter. Three went in, three came out with the dollar machine strapped to the dolly and off the property we went.

Wheeling this thing right down the road at

About one thirty a.m. and trying to be quiet.

That was a joke all in its self. You could hear us coming no doubt. One of our friends woke up and came outside; that's how quiet we were.

We kept going though and made it to a friend's house. At his house he had a tree fort out back. It was a real tree fort made by us out of scraps of wood, used nails and rope. Not an adult home in the trees.

Anyway, along the way there we busted the machine open. That was easy with hammers and crowbars. We just beat it open!

With our socks full of coins and cash we climbed into the fort and started counting our cash.

We were high on adrenaline. We were all loud and laughing. We did it! Someone threw the money in the air and shouted "We're Rich".

About that time we heard a stern voice say, "You boys come down here". We were busted.

Climbing down we saw it was one of our fathers. He talked to us about what we had done and how it was wrong.

Bottom line we bagged up the money went back up to the college and found out where the guard was. We then dropped the money bag where the machine had been sitting, Ha-Ha-Ha, He didn't want us to get caught you know.

We all went back home; it was about Four O'clock in the morning. But wait; one of the guys wasn't home thirty minutes. He's up and after the money. He gets it and the next morning guess what; Party Time!

You know we got the cash. We kept eating there as well. They replaced the machine. This time they chained it to the wall.

REALLY?

The third time we were going to get it they had welded it to plates on the floor. Each leg had a two inch bead I explained. My father had a sharp chisel. I had seen him go right through steel with it.

We all just looked at each other. The thrill was gone. It was just about the money now and that's kind of like steeling.

We didn't need the money that bad and we were onto another adventure; Never to go on the college property again.

There were some real good things we did in that time too. I also did some of these things by myself. I always seemed to be right there.

I don't know how to explain it but I know people who were a lot older than me. They never had to. I think some of it had to do with being a boy scout.

Let me explain.

During the time my mother and father were

Arguing and fighting they tried going to church. That's another story, but at the church they had a boy scout troop.

They enrolled me and it was great! I really enjoyed it. We went camping and the meetings were great.

They got me away from the arguing and stress of it all.

During my time in the scouts I earned a few merit badges but the one I studied the hardest for and enjoyed the most was the first aid merit badge.

They showed us CPR on a plastic man. Pressure points, they were all marked on him. It was visual, you knew right where they were.

The guy teaching it seemed very smart so I listened to every word he said. Then I memorized them plus the book he gave me.

When test time came he was shocked or

AMAZED, I DON'T KNOW WHICH ONE BUT WE HAD TO SHOW ON THE PLASTIC MAN EVERYTHING WE LEARNED. I MISSED NOTHING.

EVERYTHING HE SAID DURING CLASS I REPEATED TO HIM IN THE ORDER HE HAD TAUGHT US. THE BADGE WAS MINE WITH A PAT ON THE BACK.

HE ASKED ME "DO YOU THINK YOU ARE PREPARED TO USE IT IF NECESSARY"?

STANDING THERE FROZEN IN FRONT OF HIM, I COULD NOT ANSWER.

HE TOLD ME "IT'S OKAY; YOU HAVE THE KNOWLEDGE IF EVER THE TIME COMES". AND OH: IT WASN'T LONG".

THE FIRST TIME I USED WHAT I HAD LEARNED I WAS STILL A SCOUT. SUMMER TIME, IT WAS AFTERNOON;

ME AND ONE OF THE BOYS WERE COMING BACK FROM THE STORE WITH A SODA POP. WE HAD BEEN MOWING GRASS ALL MORNING LONG.

I SAW THIS MAN RIDING A BIKE TOWARD US WHEN

He kinda collapsed off the bike and ended up in this ditch full of water.

We ran up to him. He was face first in the ditch and shaking.

We jumped in pulling him out I could see his face was covered in mud. We drug him up the bank and placed him on his side. We wiped the mud and goo off his face.

He seemed to be breathing okay to me but he had his mouth closed tight and he was jerking and shaking.

The first thing I thought of was that he was having an epileptic seizure. Back then they said you should put something in between their teeth to keep them from biting their tongue; That wasn't happening.

I stayed there with him and my friend ran over to the closest house and told them to call an ambulance.

By the time the ambulance showed up he

HAD STOPPED SHAKING BUT WAS STILL NOT ALL TOGETHER.

THEY TOOK HIM TO THE HOSPITAL BUT NOT BEFORE GETTING OUR NAMES AND ADDRESSES.

THOSE GUYS TOLD US HOW GOOD OF A JOB WE DID. WE SHOULD BE PROUD OF OUR SELVES BECAUSE IF WE HADN'T OF BEEN THERE THE MAN MIGHT OF DROWN IN THAT DITCH.

ABOUT A WEEK LATER THAT GUY SHOWED UP AT MY HOUSE AND THANKED ME. HE TALKED TO MY MOTHER FOR A MINUTE THEN LEFT.

I ASKED MY MOTHER WHAT HE HAD SAID. SHE SMILED AND SAID, "HE SAID YOU WERE A SPECIAL PERSON AND I TOLD THE MAN WHO TAUGHT THE FIRST AID CLASS.

"WHAT HAPPENED"

"HE JUST SMILED AND SAID YOU JUST ANSWERED MY QUESTION, YOU DID A GOOD JOB AND THAT IT'S IMPORTANT TO STAY PREPARED TO HELP".

THERE HAVE BEEN MANY TIMES SINCE THEN THAT

I HAVE BEEN THERE FOR SOMEONE WHEN THEY NEEDED SOMEONE.

I WILL TELL YOU ONE MORE NOW THEN MOVE ON BUT I WOULD LIKE TO SAY IT'S BECAUSE OF SOMEONE IN THE BOY SCOUTS WHO TOOK THE TIME TO TEACH ME THE BASICS THE RIGHT WAY.

FROM THAT POINT ON IF A BUSINESS THAT I WAS WORKING FOR OFFERED A FIRST AID COURSE, I WAS THERE. EVERYONE SHOULD KNOW SOME, A LITTLE PREPARED IS BETTER THAN NONE.

ANYWAY; I WAS A MANAGER FOR A GARAGE AND I TOLD ONE OF THE MECHANICS TO GO CHANGE A BATTERY IN A CAR THAT WAS OUT BESIDES THE BUILDING.

I HAD CHECKED EARLIER, IT WAS BAD AND THE TERMINALS WERE CORRODED. I PARKED IT THERE BECAUSE IT WOULDN'T TAKE LONG TO CLEAN THE TERMINALS AND CHANGE THE BATTERY.

THE SUN WAS HIGH IN THE SKY AND THIS GUY WANTED TO PUT IT IN THE BAY IN THE SHADE. SO HE TOOK THE JUMP CART OUT TO JUMP IT AND WHEN

HE DID THE BATTERY EXPLODED. (THEY WILL DO THAT) I HEARD IT AND KNEW RIGHT AWAY WHAT HAD HAPPENED.

WHEN I GOT THERE HE HAD HIS HANDS OVER HIS FACE. HIS SHIRT WAS WET WITH ACID. THE OTHER MECHANIC WAS RUNNING FOR THE WATER HOSE.

THERE WAS A CONTAINER FULL OF WATER IN FRONT OF THE BAY JUST DOWN FROM THE CAR.

WE USED IT TO CHECK FOR AIR LEAKS IN TIRES. IT WAS A LITTLE DIRTY BUT WATER AS FAST AS YOU CAN WITH ACID IN YOUR EYES. BOTTOM LINE!

SO I GRABBED THIS GUY BY HIS ARMS AND BELT; YOU GUESSED IT; IN THE TIRE CHECKING CONTAINER HE WENT HEAD FIRST.

HE STAYED DOWN FOR A MINUTE RUBBING HIS FACE WHEN HE RAISED UP TO GET A BREATH. I RIPPED HIS SHIRT OFF HIM THEN STUCK HIS HEAD BACK IN.

WHEN HE RAISED UP AGAIN THE GUY WITH THE WATER HOSE WAS BACK THERE AND SOMEONE ELSE

HAD BROUGHT THE EYE FLUSH KIT.

WE WORKED ON HIM UNTIL THE AMBULANCE SHOWED UP.

HE CAME BACK TO WORK TWO DAYS LATER. HIS FACE WAS RED AND RAW BUT HE COULD SEE.

HE TOLD ME THE DOCTOR MADE IT CLEAR ON HOW LUCKY HE WAS THAT HE'D GOTTEN HIS EYES FLUSHED SO FAST.

SO HE TOLD ME THANKS. HE ALSO ADDED WE SHOULD CLEAN THE TIRE CONTAINER MORE. WE LAUGHED THEN WENT ABOUT THE DAY.

WELL DURING THE TIME I LIVED IN THE EFFICIENCY AND MY FATHER WAS LOSING HIS MIND AGAIN I BECAME A CERTIFIED MECHANIC, STARTED PLAYING WITH DIRT BIKES AND DID A LOT OF FISHING.

I ALSO DRANK A LOT OF BEER AND DID THINGS THAT GO ALONG WITH IT.

THEN ONE DAY MY FATHER INTRODUCED ME TO HIS THIRD WIFE TO BE. SHE WAS A BARMAID AND A LITTLE OLDER THAN HIM. SHE WAS A "BOTTLE

BLOND" AND NOT BAD LOOKING.

THE FUNNY THING WAS THOUGH THAT SHE DIDN'T DRINK, NOT A DROP. I FOUND OUT AN HOUR LATER THAT SHE WAS A POT SMOKER. SO WE ALL SMOKED A LITTLE AND TALKED.

I FOUND OUT THAT SHE HAD A SON AND A DAUGHTER. SHE TALKED ABOUT HER LIFE AS WE HUNG OUT THERE FOR ABOUT FOUR OR FIVE HOURS.

IN THAT TIME I NOTICED MY FATHER HAD NOTHING TO DRINK BUT A COKE.

IT WAS NICE TO SEE HIM SOBER OR WHATEVER YOU WANT TO CALL IT. NO ALCOHOL WAS GOOD FOR HIM. HIS VOICE SOUNDED STRONG AND HE LOOKED GOOD.

I THOUGHT TO MYSELF THAT MAYBE HE WAS FINALLY HAPPY. LATER I KNEW HE WAS.

ANYWAY, I LEFT AND WENT ON WORKING AND DOING MY THING. LATER THEY MOVED INTO AN APARTMENT TOGETHER. HE WAS WORKING STEADY; EVERYTHING SEEMED GOOD.

Until one day she was driving home from work and someone had fallen asleep or something and hit her in her car.

It did a lot of damage to the car and hurt her, not real bad but bad enough to keep her out of work for a while.

In that time these two people hatched a plan. They knew she was going to get some money from the wreck; Ten or Fifteen Thousand Dollars. When the dust finally settled they got her money.

That weekend they called us all over to their apartment. She had chips, pizza, soda, fruit but no alcohol was to be found.

There was a bag of weed on the table though with papers and a couple of joints already rolled.

We smoked some and sat around eating and talking. Then my dad said, "We have something to tell all of you".

Life was about change for all of us. The plan was to:

Wait a minute; I am just going to write the story because what he told us that day; those two people did!

Without missing a beat they moved to a town about an hour away, got a little apartment then set out on their dream.

They had a trailer full of furniture, clothing, pots and pans etcetera, etcetera.

Nobody knew them so they took the things in the trailer to the swap meet (flea market or park-n-swap, whatever you want to call it) and they rented a booth.

They didn't just work there Friday, Saturday and Sunday. They were in and out of it seven days a week buying, selling and trading anything of value.

It didn't take them long to be known around there either.

They went from a booth to five. It was like a carnival when they were working. People were flocked around there booths and everyone wanted their attention.

Her daughter and son-in-law had moved down there. They were working with them.

They were not making a lot of money but they were making enough to put away a hundred or two a week, plus he made a lot of connections.

Most people don't realize the types of people who go to these places;

Doctors, Lawyers, Car salesman, Grocery clerks, Mechanics; On and on! You never know!

Some people just collect different things and they hunt them in all kinds of places.

They were there for a little over a year when they had met someone who had an old run down car lot. On that car lot there

WERE ABOUT TEN CARS, NONE OF WHICH RAN.

The grass was high and needed a good mowing. There was an old house out back. It had not been lived in for years.

He made a deal with this old man then gave him all the cash but a Thousand Dollars.

My dad told me the old man made the deal, he couldn't work the car lot any more but wanted a good game to watch.

The old man had been a player in his time and if my dad and his wife failed or succeeded, the old man had something to do. It was good for everyone.

They kept the booths at the Park-n-Swap but the real game was the car lot. They cleaned up the lot and the house. The kids worked the booths.

They all moved into the house and my father got some of the old cars running again.

They sold them on payments. He would take anything of value in trade which he would set out on tables he had built from junk wood and things that he had salvaged.

It wasn't long before he let the booths at the flea market go because the happening place was the car lot. There were people there from the time they opened until midnight and beyond.

They were bringing in payments, stuff to sell, cars boats, motorcycles; you name it they were selling it.

Don't miss a payment, he would send somebody to collect. He knew a lot of, shall we say, interesting people and if they couldn't collect he would and that you could bet your life on.

He had no fear!

They were doing good when he called me one day and told me to come down for the weekend. He wanted to talk to me. I said okay;

Of course I could hear by the sound of his voice something exciting was going to happen.

I would have left right then if he would have asked. The weekend rolled around and I went down to the old house where he met me.

His wife and the kids were there. It was as usual, snacks, soda and of course a bag pot on the coffee table with papers plus a few pre-rolls.

He laid the plan out for me in front of everyone there. Then he said in front of everyone that he needed to talk to me; just me.

That made me a little nervous but we walked outside and this is what he said.

"Son, you know me and you should be able to see and hear that this is one of those

GIFTS; THE KIND THAT VERY FEW PEOPLE GET.

You know that I am a believer and now is the time to live with no fear, it's the right time and were in the right place" looking into my eyes as I nodded and agreed, he continued.

"It can be a time of happiness and fun. It will be a time of hard work also. I will be counting on you a lot. We need a lot of cars fixed so they can be sold."

After a few questions he informed me that his wife's son would be moving down and would be helping me, although he didn't know much about fixing cars; him and I would live in the old house behind the car lot.

It wouldn't be your average eight hours a day with weekends off, that it would be work till you drop, sleep and do it again. It would be like this until it rolled into the next level of their plan.

I was alright with all this but knew that there was one more thing that I had to do before it could happen;

I had to quit drinking for a while so I added that smoking pot would be okay. We laughed.

It was a Saturday afternoon and I was down there working by Monday morning.

I stayed in the screened in porch for about a week or two while they moved. It was summer time so it was no big deal.

My boss who had been so good to me was pissed but there was no way I was going to miss this adventure and what an adventure it turned out to be.

Plus my boss was my fathers and my friend. He knew what was happening so he got over it quick.

Her son came down and we lived in the old house as we worked our butts off fixing cars. We fixed a lot and my father sold them as fast as we could fix em. It was sell one and buy two or three not running.

Soon we had other people helping us just as he had explained that day but first let me tell you a few funny stories before I move on to the next level of his plan.

He gathered us all up, it was a Sunday and the usual atmosphere. Then he began to tell us his father was coming down the following Saturday.

We had to move all the cars off the lot. He was bringing in truck loads of white rock to cover the dirt.

He wanted everything painted or cleaned and it would be our grand opening. He wanted his father to see the game plus he wanted to build a cage in the trees.

We then all at the same time said "For What",

"White Rabbits", he replied. "We are going to have white tree rabbits for our grand opening".

We all started laughing then he said "I have a surprise for you son" and pointed his finger at me.

I asked him what it was and he told me that he would let me know on Friday.

That Monday we set off working on cleaning the lot and building the cage in the tree. It was big and had two levels. We were done with it Wednesday afternoon.

That's when we brought two big white rabbits to the lot and he put them in the cage.

We fed them and played with them until Friday morning when he called me into his office. This is what he said;

"Go to the store and buy the most colorful loud shirt you can buy. It has to be a button

UP TYPE. THEN FIND A HAT, SOMETHING THAT WILL STAND OUT. ALSO GET YOU SOME CRAZY SUNGLASSES THEN COME BACK AND SEE ME".

I GUESS MY BIG BLACK BOOTS AND BLUE JEANS WHERE NOT CRAZY ENOUGH ALONG WITH MY BRIGHT RED HAIR.

WITH THE MISSION HE HAD SENT ME ON ACCOMPLISHED I RETURNED AND TO MY SURPRISE HE HAD PUT THE RABBITS IN CAT BODY COLLARS; ORANGE AND BLUE WITH MATCHING LEASHES.

HE EXPLAINED TO ME HOW HE WANTED ME TO WALK THE RABBITS UP AND DOWN THE SIDE WALK IN FRONT OF THE CAR LOT THAT SATURDAY DRESSED IN THE SHIRT AND HAT WITH THE GLASSES;

HE STARTED LAUGHING. THEN I STARTED, IT WAS ONE OF THOSE MOMENTS. HE KNEW I WOULD DO IT AND WITH A FLAIR.

WE BLEW THE BALLOONS UP AND PUT THE FLAGS ON THE CAR ANTENNAS. (THEY DIDN'T MAKE THEM IN THE WINDSHIELDS AT THAT POINT)

The next morning when my Grandfather showed up there I was all dressed up and walking those rabbits up and down the sidewalk waving people into the lot.

People were taking pictures and blowing their horns. We were selling cars. It was truly a moment. We still laugh about it.

My Grandfather walked up to me and smiled, said "You didn't fall far from the tree".

We had a great day. It was the last time I saw my grandfather whom I loved more than words can explain. He was real and a great warrior.

The rabbits lived in the tree cage until it was over. People would bring their kids to see them. It was an attraction to say the least.

Just keep bringing your payments! Oh yeah and some carrots for the rabbits. Ha-Ha-Ha

The car lot was more than that. It was a meeting place for family and friends because if you bought a car from my father you would definitely bring your family and your friends to buy one as well and since payments were to be paid either Friday or Saturday our customers would see their family and friends there too.

So we had picnic tables and chairs, a shaded area, barbeque and a little fire pit. We would cook hamburgers, hot dogs and do corn on the cob.

None of the people were wealthy as far as cash goes but some of them were wealthy with personality and dreams.

So now I write the story of "Jake The Snake". My father sold him a 73 Monte Carlo and then made a dream of his come true.

Jake worked a minimum wage job and lived in a trailer park in the rough part of town.

He spent all his spare time training to be a boxer.

He would tell my father about his dream of getting into the ring. He would show my father all his moves.

Jake was an armature but my father made him feel like a pro. Jake told my father about a fight he was going to have. It was in some arena about twenty minutes from the lot.

The game was on. He had red T-Shirts made up, they had "Jake the Snake" on the back with the day and time of the fight.

On the front it had the car lot name and a statement that read "Proud Sponsor Of Jake The Snake".

He gave them out to everyone who came to the car lot, plus random people up and down the street. He carried twenty or thirty with him at all times.

He even had posters of him made and hung

them around the office and the front door. He also took a few down to where we had lunch quite often.

Jake became the man to know. Everyone was shaking his hand and wishing him good luck on the fight. You could watch the change in this man happen.

Jake had his chin up, he had confidence. There was a girl who started hanging out with him. Life was good but it was about to get better.

The night of the fight, my father and his wife took Jake to the arena in their Cadillac Eldorado Convertible.

He was in the back seat as they were chauffeuring him to the fight. They had the t-shirts on with white pants, gold sunglasses and they both had red hats on.

Oh yeah; I was following them in a white Chevy station wagon. There were nine of us in the wagon. We all had the t-shirts on.

We walked him in so he could get ready. There were a few more people inside with the shirts on. They started yelling, "Jake The Snake", "Jake The Snake" over and over as he walked by.

Other people turned to see what was going on. Jake told my dad he felt like he was on top of the world at that moment.

He went into his room to get ready for the fight and about thirty minutes later he came out. For his fight more people had showed up.

There must have been thirty or forty people there for him. They were all chanting "Jake The Snake" and so was I.

He was knocked out in the second. It didn't quite turn out the way we all wanted but it was good for all of us.

We sold more cars, met more people and as for Jake; He got a girlfriend out of it, plus a memory I am sure he will never forget.

I have never forgotten what my father did for another human just because he could. I once told him that he should be proud of himself for what he did and that I was proud of him.

He just smiled at me and said that we all have been blessed.

So the next level was off the chart. My father moved to a house on the Island first.

We were in the money. He was buying cars from the auction house plus trading things. People were at the lot all the time.

New and old faces were coming to see what it was all about. If you walked onto the lot, you left with something.

I've seen my dad sell cars to people that they didn't even know they wanted.

For example: A man came to buy a truck, we didn't have one running at the time but I had just finished putting a Pontiac Four Hundred engine in this Sixty Nine El Camino and it was nice.

My dad walked the man around to where I was working, putting tires and craggier rims on it. He was telling the man about the car and all that we had done to it.

Then he asked me if I thought it would last. Of course I said yes, I was the one who built it.

The man then tells my dad that he didn't want an El Camino.

There was a moment: we were all standing there when the man says your horse shoe is turned wrong. My father had them hanging everywhere.

My dad asked what he had meant; the man replies, it's supposed to be turned the other way to hold the luck in. There was about 7 seconds of silence and then, "That's not what I believe" my father told the man.

"You are supposed to let the luck run out so it can be shared and that way the Lord is always filling it back up" he added.

"Today is your lucky day however!"

He told the man. "Take the El Camino and drive it for three days and if you don't like it I'll find you a truck or give you your money back".

Let me tell you, that man owned that car, he might even still have it. We also sold him his family car.

It didn't matter to my father what they were driving, it was about the payments and the people. He always said that it was his job and ours to keep them riding as long as they were making their payments.

If your car broke down, no problem, bring it in and if he couldn't fix it then we did. If not we gave you another one, same price just for you. They all had warranties, cars, boats, and motorcycles whatever it was he

HAD SOLD.

People would come in talking about buying a boat or whatever and he would tell them, "Take mine, see if you like it". They would look at him funny. "It needs to be run anyway" he would tell them.

My dad's wife's daughter and son-in-law where next to move to the Island; and at this point it really became a show. I can't explain it; It was like an outer body experience, if you have ever had one and I'm Not just talking about life.

My dad had met some people who were one thing during the day and another during the night or when they had time away from what they did during the day.

That's all I am aloud to say; My dad told me not to go into detail about them.

Anyway they paid to be a part of it all. He knew a lot of people like I said. He would put these people in touch with other people who were like them. People knew who we were.

It come time to be my turn to move to the Island. His wife's son and I had a new apartment on the beach. It was up on poles.

One apartment faced the Gulf and the other faced the Inter Coastal Water Way. Two girls lived in that one. My bedroom door was glass and it opened onto a deck that had a million dollar view.

We drove whatever we wanted from off the lot. Convertibles and Motorcycles were my choice but my father stopped me from riding the motorcycles and I know why.

I would take them on the beach. Generally just raise cane with them. Sand was not a good thing for his B.S.A. Chopper, totally custom.

We didn't have to go into work as early anymore, he had sales people. I would get up, jump in the 68 GTO Convertible I was driving and drive down the beach a ways and go

Snorkeling in the mornings. Then I would go to work afterwards.

I would come home and party till when ever. The girls who lived in the other apartment were waitresses on the beach and they knew people.

There were people around there all the time. I could write story after story about all the parties. Here's one for you.

A friend of mine that I hadn't seen in a while came down on a Friday. He got there late and the party that night had already mellowed out.

So we sat around. He drank some beer, I still wasn't drinking and I think that puzzled him but I just rolled one up and drank a soda pop.

The next day we went to the beach and then to the restaurant/bar that the girls were working at. I introduced my friend to the girls then told them we needed to have a big

PARTY.

About five o'clock or so that day there must have been twenty five or thirty people there at the apartments. The party was on.

Someone had backed there truck up to the balcony. The bed of their truck had a bed mattress in it. People were jumping off the balcony into it.

The radio was loud and people were playing in the water hose. Some had clothes on. We were all having a good time.

It was about eleven or so at night and some slope head try's to climb the power box pole. It snaps off the side of the building, sparks are flying everywhere.

Well; there goes the power. The party was over. My friend went home that Sunday. All he said was, you're crazy and thank you for the good time.

There is one other time I will write about

ON THE BEACH BEFORE I MOVE ON TO THE TRAGEDY. IT'S ABOUT WHO YOU KNOW SOMETIMES.

ME AND A FRIEND OF MINE WENT TO THIS BEACH BAR. THEY HAD A LIVE BAND PLAYING. WE WERE PLAYING VOLLEY BALL AND DANCING, JUST HAVING A GOOD TIME. IT GOT DARK WHILE WE WERE THERE.

WE WERE OUT OF PRE-ROLL'S SO WE WENT OUT TO THE CAR I WAS DRIVING. IT WAS A MERCURY CONVERTIBLE, DON'T REMEMBER WHAT KIND BUT IT WAS HUGE AND YELLOW AND THE TOP WAS DOWN.

I REACHED UNDER THE FRONT DRIVER'S SEAT AND PULLED OUT A FOUR FINGER BAG AND STARTED ROLLING ONE.

WHEN I FINISHED I PUT THE BAG ON THE FRONT SEAT THEN I LIT IT. WE WERE SMOKING WHEN A COP WALKS UP.

"WHAT'S UP" HE SAYS, WITH A JOINT HANGING OUT OF MY FRIENDS MOUTH AND HE SEES THE BAG LAYING THERE, HE TELLS MY FRIEND TO PUT THE joint out.

He starts in with his B.S. on how it's illegal. He went on and on then calls for back up. After a few minutes another cop pulls up.

They started talking when the one cop that caught us gets in his car and drives off. His back up walks up and tells me to hand him the bag. I do!

While all this is going on all I could think about was how much jail time I was going to get because there were three more bags under the seat. Then I flashed back to reality as the cop said;

"Today's you lucky day, I know your father and I owe him one. I will call him later and tell him we're even".

With a sigh, he told us to get out of there and to go home. Then with a "Call it a night, Understand" he walked away.

I started the car up and headed out. I looked at Dave then asked him where he wanted to go. The look on his face made me start laughing. He asked me if I had lost my mind.

"Take me home" he said "Dint you hear that cop". I took him home then went to a party.

A couple of days later my father said to me, "I heard that you were lucky the other night. You know I DON'T know everyone".

In the mean time my father bought a 28 foot yacht. He parked it in the canal that led to the bay behind his house. He was proud of his boat.

Living on the Island was a young man's dream if you ask me. That's if you like the beach, Convertibles, Motorcycles, girls and parties. The money, I almost forgot about that.

It went this way for a time. It was like a dream when I look back at it. A good one

THAT YOU NEVER WANTED TO END AND THEN YOU WAKE UP, YOU TRY TO GO BACK TO SLEEP BUT CANT.

I DON'T KNOW HOW ELSE TO EXPLAIN IT. IT WAS NOTHING LIKE REALITY TV. THERE IS NO WAY TO FILM IT, IT'S REAL. NO SECOND TAKES. YOU CAN'T DELETE THE PARTS YOU DON'T LIKE.

HEAR IT COMES THE PART I DON'T LIKE; MY FATHER PULLS INTO THE CAR LOT ONE DAY, WALKS OVER TO ME AND SAYS "LET'S GO FOR A RIDE".

"OKAY, WHAT'S UP" I ASK. HE DOESN'T SAY ANYTHING. I GET IN AND WE DRIVE. WE WERE ABOUT TWO MILES AWAY FROM THE LOT WHEN HE TELLS ME THAT HIS WIFE HAS CANCER AND SHE HAS A SHORT TIME TO LIVE.

IF YOU HAVE READ THIS BOOK THIS FAR THAN YOU KNOW. **DEVISTATION, HEARTBREAK, INSANETY** ALL OF IT CAME IN. JUST LIKE THAT, INSTANTLY, IT WAS ALL OVER AND HE KNEW IT.

HE TRIED TO HOLD IT TOGETHER BUT IT ALL STARTED SLIPPING AWAY.

She was sick and he was there for her. Nothing else mattered. I was no help at all. After he had told me I started drinking. I couldn't believe this was happening to him again.

We all moved off the Island. We tried to keep the lot going but it was not in the cards.

Me personally; I lost it. I got a drunken driving ticket. The story doesn't matter right now but it was insanity at its best. I got a job working at the garbage company because money was short.

That was no big deal, they paid me well and I could fix anything they had but I wanted to be a slinger.

They liked me there, so when they needed one, they would let me do it. We were done with our routes by noon. People left beer in coolers for us.

We stopped by the beach for lunch and

drank the beer of course. Good job; Loved it. Once you get past the smell, the parties on.

I met some people there. They were different. We hung out but it wasn't enough for me. I was watching my father fall apart again.

I should have left when I got the ticket but it took one more screw up.

It was about two or three in the afternoon. Me, my dads' wife's daughter and her son-in-law took my dad's yacht out. It was a Saturday and we were already half drunk.

We took a case of beer and a gallon of rum with some punch. We headed out for an Island where everyone partied.

Getting there was no problem. We all partied till dark and had a great time. We pulled the anchor up and headed in.

Along the way we ran out of beer, yet no

problem. There was a bait shop on a dock they also sold beer. We headed there first.

Well; We hit the dock. The man ran outside screaming What Happened. He ran up to where we were. He knew who we were and he saw we were drunk. He told us to leave but I told him to give us a case of beer.

He told us No the first time but I wouldn't quit. He finally gave in. Oh Yeah; when we hit the dock it knocked a hole in my dad's boat just above the water line, no problem.

So across the bay we went and up the canal to my dad house we went. Radio was wide open, music jammin. You could hear us coming. People were turning their lights on, dogs were barking.

My father was waiting for us in his underwear. He was mad to say the least.

When we started to dock I jumped off and ran for my car. I wasn't staying around. He can be mean when he's mad. I mean, Scary mean.

On the way home the police stopped me. They let me go for being drunk but they ticked me for reckless driving

I didn't care, I drove over to the brigade, stood there, ripped the ticket up and threw it off. Then I went about my business.

My father gave up being mad about his boat then one day the police showed up at the car lot. What was left of it anyways!

They were looking for me. Having a bench warrant for my arrest; Yeah: Remember the ticket I ripped up, my father had no choice but to tell them where I worked.

As soon as they left he called my boss. My boss knew him, heck the garbage man knows everybody.

He also knows when you have money or not. He knows when you party and how. He knows what you drink; it's all in your

Garbage cans.

I was on route when the boss radioed the driver. The driver told me that the cops were on their way. That's when I said See ya'll Later!

I went to a friend's house and he gave me a ride to my car.

I went home and packed my things then headed for the car lot to talk to my dad. We sat under that tree where the rabbits were and talked.

I told him that I was going back to Arizona. He told me that it sounded like a good idea compared to jail.

So with my Ninety Three Dollars, a green and white Caprice 2 door 73 model, I headed for Arizona for a thousand stories to gather and people to meet. That's where I met Charlie.

Oh yeah; Charlie; Let me get back to him!

I had been driving for a while, I don't really remember a lot of what I saw or even the speed. I had been on cruise control in my head not the car.

You know if you've been driving too long when it becomes;

Was that light red or green, yeah the one I just went through or; When you pass your friends drive way and realize it as you passed it.

I then knew I was zoned out but somehow I made it to Charlie's town.

I pulled into the gas station, filled up and asked for directions. I am not proud, forget the GPS and talk to a local, they know. She told me about where he lived so I headed that way.

Turning onto the street he lived on I was looking for a number that was funny to me.

When I saw his house there was an old

TRUCK, A VAN AND A PILE OF LUMBER OUT FRONT. THERE WAS ALSO A LOT OF MISCELLANEOUS STUFF, ONCE A COLLECTOR ALWAYS A COLLECTOR.

THEN I SAW HIM, BENT OVER A LAWN MOWER WITH A BEER IN HIS HAND. HE SAW ME ABOUT THE SAME TIME HE STARTED WALKING TOWARDS ME WITH A SMILE ON HIS FACE.

I COULD SEE HE HAD TWO BEERS IN HIS HAND. "SAME OLD CHARLIE; SO FAR", HE HAD GRABBED ONE FOR ME.

HE GAVE ME A HUG, ASKED ABOUT MY TRIP THEN SAID, "I MISSED YOU, GLAD YOU'RE HERE. LET'S HAVE A BEER".

SO WE WALKED OVER TO A TABLE WITH SOME CHAIRS AROUND IT UNDERNEATH A SHADE TREE OF COURSE.

WE SAT THERE FOR HOURS TALKING AND DRINKING BEER. NOBODY ELSE WAS AROUND. HE FIRED UP THE GRILL, NOTHING FANCY, HAMBURGERS ON BREAD, MUSTARD OR KETCHUP OR BOTH, YOUR CHOICE AND THAT'S IT. IT WAS FINE with me, no mess.

After we ate I asked him where everyone was. He told me that he lived there by himself for now. His girlfriend had gotten locked up for some bull ship.

They gave her thirty days and that she would be out in a couple of days, no big deal. I thought Okay, didn't bother me, why should it. He tried to explain but I really wasn't listening that well.

I could tell something wasn't right. We kept talking and drinking. It was late in the evening when I finally made it to his front porch and into his house.

I was tired so I didn't notice much about his place. I just went to bed. He pointed to a room and that's where I went. I didn't even grab my stuff out of the car.

The next morning we got up, he grabbed a beer and a coffee for me. He had sugar but that was it. Okay with me. Then he said "Let's

GO DOWN TO THE RIVER".

There would be people he knew down there. "We could swim and have a few beers" and he would introduce me to some people he knew. It was just where the locals hang out you know.

I replied with "Yea" and we jumped in his old truck and headed out.

First stop was a little general store. We needed supplies; beer, two twelve packs, a cup of coffee, a gallon of water, one bag of chips, pack of hot dogs, loaf of bread and don't forget the cigarettes we were out of.

Charlie cracks his second beer and were on our way. It wasn't far. I still had a little coffee left when we arrived. We stepped out and Charlie opens another beer.

Walking down to the edge of the river, I just stopped and looked at it for a while. Seeing the little fish move around made me smile.

Charlie said "Let's jump in". It was a little cool but not cold, It was nice. Charlie had brought a twelve pack with him. You ready, "Yes" I replied.

We sat there and some friends of his came by but they were headed to a different part of the river. They asked us if we wanted to go but we were good right where we were at.

We had a few more beers. He was talking about our time in Arizona a lot. I asked him if everything was okay with him. He said yes; that he was working and wasn't making a lot of money.

He didn't need much, his health was good and it was just the normal ship in life that was getting to him a little. He said that he had been thinking of me. He had been wondering if we would ever share another beer and laugh together. It had been a few years.

Then he said "you'll like my girlfriend, she's crazy but in a good way. When she gets out we can go check the people out that I told you about". He also mentioned that he told her a lot of stories about him and I in Arizona.

Before I go any further I am going to have to tell you about Arizona. That's going to have to happen in another book. How I got there and who I met. When I met Charlie and How!

I will tell you; you think the stories so far are entertaining. You haven't read anything yet. Some you won't believe but there all true. I didn't just make them up. I was there 95% of the time.

I do hope you understand a little about who we are, Charlie and I, because our personalities together made for a lot of fun. Hard work was a big part of it as well. Our ups were great and our downs painful at times.

I don't mean to sell you. You know the first part of our lives; it's up to you to read the next book or not. If you have read this book and its entirety I would like to say;

THANK YOU !

Dictionary:

<u>Alabama Fire-Log</u>: Old used tire: Usually burned at night. In daylight someone probably needs help; Another words Rescued or needs the other type of help: <u>How to say it</u>; Ship were out of wood, Oh it's alright, Hell I got a Alabama Fire-Log out back.

<u>Bit</u>: Time: Usually used when you have no Idea how long you will be gone. But you do intend to come back. <u>How to say it</u>: I am going to the beer store I'll be back in a bit.

<u>Booger Light</u>: A Light; Used to light up real dark places that might seem a little scary at night or so you can see where you're going and nobody can walk up on you without being seen. <u>How to say it</u>; Boy go out to the shed and grab me the hammer, the booger lights on; there ain't nothin going to get you, Now Go.

<u>Boot Legging</u>: Illegal Alcohol sales; Usually preformed in counties that don't sell Alcohol on Sundays or at all; Usually done to make some easy money: If caught you will go to jail. <u>How it's said</u>:

My head hurts, I need a beer. Your friend says' you drank it all last night and you know they don't sell it today; It's Sunday. You reply with, Then let's go to the boot legers.

<u>Corn Binder</u>: Equipment made by International; Usually found doing Heavy work, Tractors and Semi-Truck. They made a pick-up truck also along with other things; <u>How to say it:</u> your friend says, My equipment or truck is stuck, you just yell at your kid; Go around to the barn and get that old corn binder out back and jerk him out of the mud.

<u>Escapade</u>: Wild adventure; Usually involves other people; <u>How it was said to me by my Father</u>: There was nothing wrong with that truck. You and Lonnie burned it and hauled it for scrap. I am sick and tired of your escapades you little ship.

<u>Lemans'</u>: A car made by Pontiac. They were cool until the late Seventies. <u>How to say it</u>: If we take that six-cylinder out of that Lemans', put a V8 in it and put the G.T.O stuff on it from that wrecked on they will never know the difference. (That's where "Clone" cars came from)

<u>Nail Head</u>: A V8 Motor made by General Motors. Usually found in Buicks. <u>How to say it</u>: You say, I told her it will take most of the day to change the

VALVE COVERS ON THE CAR, YOUR FRIEND SAYS, THAT'S FUNNY IT HAS A NAIL HEAD IN IT. YOU SAY, JUST GO GET THE BEER AND POLES I'LL BE DONE IN TWENTY MINUTES.

NEY SAYERS: IT CAN'T BE DONE. SPOKE BY PEOPLE WHO JUST DON'T GET IT. HOW IT'S BEEN SAID TO ME: YOU CAN'T TAKE THAT OLD TRUCK AND TURN IT INTO A TOW TRUCK THAT'S USEABLE. I JUST DID IT. NO USE TO EXPLAIN.

PRE-ROLLS: HAND ROLLED CIGARETTE; USUALLY DONE PRIOR TO A PARTY OR AN EVENT SO YOU DON'T HAVE TO STOP WHAT YOU'RE DOING AND ROLL ONE. HOW IT'S SAID: YOU'RE SO THOUGHTFUL, YOU HAVE PRE-ROLLS. BY THE WAY THEY CAN HAVE MORE THAN TOBACCO IN THEM; MARIJUANA FOR EXAMPLE.

PYLONS: STRUCTURES; CAN BE MADE OF CONCRETE AND OTHER MATERIAL. HOW IT'S SAID: MAN THOSE PYLONS ARE HUGE AND THAT BRIDGE THEIR HOLDING UP IS TOO.

RAUCOUS: HARSH, JARRING; USUALLY SPOKE WHEN CAUGHT OFF GUARD; HOW IT'S SPOKEN: I HAVE NO IDEA. I'VE USED IT ONCE BECAUSE I HEARD SOMEONE ELSE USE IT. HA-HA!

RAVAGED: DEVASTATION; USED TO DESCRIBE HURRICANE CAMILLE. NOTHING REALLY MORE TO SAY.

SLED: AN OLD USE VEHICLE, CAR OR TRUCK. USUALLY YOU HAVE LITTLE MONEY INTO IT. IT MAY NOT LOOK SO GOOD

BUT RUNS GREAT. THE TIRES HOLD AIR AND WILL GET YOU TO POINT A AND BACK WITHOUT CALLING YOUR FRIENDS TO COME AND GET YOU. <u>HOW IT'S SAID</u>: OH IT'S JUST AN OLD SLED BUT IT GETS ME AROUND.

<u>SLOPE HEAD</u>: A PERSON; USUALLY BIG AND NOT TO BRIGHT. <u>HOW IT'S USUALLY USED</u>: SOME BIG MUSCLE BOUND SLOPE HEAD WAS TRYING TO START SOMETHING WITH ME BUT HE WASN'T THAT SMART. I HIT HIM WITH THE HAMMER I WAS USING TO FIX THE FENCE AT THAT TIME.

<u>STOVED UP</u>: CAN'T MOVE. USUALLY USED WHEN YOUR BONES HURT OR MUSCLES. <u>HOW IT'S SAID</u>: I CAN'T BEND OVER AND PICK UP THAT SCREW DRIVER, I AM ALL STOVED UP. YOU DO IT.

<u>STUMP KNOCKERS</u>: FISH; VERY COLORFUL NOT REAL BIG. USUALLY USED TO DESCRIBE SMALL FISH IN THE SOUTH.. <u>HOW IT'S USED</u>: WE CAUGHT A BUNCH OF STUMP KNOCKERS AND DEEP FRIED THEM. THEY TASTED GREAT WITH THAT "COLD BEER".

Made in United States
Orlando, FL
13 April 2024